S0-BSZ-451

Biomedical Science and Technology

Recent Developments in the Pharmaceutical
and Medical Sciences

Biomedical Science and Technology

Recent Developments in the Pharmaceutical
and Medical Sciences

Edited by

A. Atilla Hıncal and
H. Süheyla Kaş

Hacettepe University
Ankara, Turkey

Plenum Press • New York and London

Library of Congress Cataloging in Publication Data

Biomedical science and technology: recent developments in the pharmaceutical and medical sciences / edited by A. Atilla Hıncal and H. Süheyla Kaş.
    p.   cm.
    "Proceedings of the Fourth International Symposium on Biomedical Science and Technology, held September 15–17, 1997, in Istanbul, Turkey"—T.p. verso.
    Includes bibliographical references and index.
    ISBN 0-306-45837-3
    1. Biomedical engineering—Congresses. 2. Medical technology—Congresses. 3. Pharmaceutical technology—Congresses. 4. Biotechnology—Congresses. I. Hıncal, A. Atilla. II. Kaş, H. Süheyla. III. International Symposium on Biomedical Science and Technology (4th: 1997: Istanbul, Turkey)
    [DNLM: 1. Technology, Pharmaceutical congresses. 2. Technology, Medical congresses. QV 778 B6155 1998]
    R856.A2B577   1998
    610′.28—dc21
DNLM/DLC
for Library of Congress
                                98-22345
                                CIP

Proceedings of the Fourth International Symposium on Biomedical Science and Technology, held September 15–17, 1997, in Istanbul, Turkey

ISBN 0-306-45837-3

© 1998 Plenum Press, New York
A Division of Plenum Publishing Corporation
233 Spring Street, New York, N.Y. 10013

http://www.plenum.com

10 9 8 7 6 5 4 3 2 1

All rights reserved

No part of this book may be reproduced, stored in a retrieval system, or transmitted in any form or by any means, electronic, mechanical, photocopying, microfilming, recording, or otherwise, without written permission from the Publisher

Printed in the United States of America

BIOMED-4

The Fourth International Symposium on Biomedical Science and Technology

SYMPOSIUM COORDINATORS

E. Pişkin   Hacettepe University
V. Hasırcı   Middle East Technical University

INTERNATIONAL SCIENTIFIC COMMITTEE

U. Akbulut	TR	H. S. Kaş	TR
G. Alaeddinoğlu	TR	R. Langer	USA
S. Benita	IL	T. Nagai	J
G. Gregoriadis	UK	E. Pişkin	TR
M. J. Groves	USA	D. Poncelet	F
N. Hasırcı	TR	Y. P. Tan	TR
V. Hasırcı	TR	A. Tümer	TR
J. Heller	USA	Y. Ulcay	TR
A. A. Hıncal	TR	D. L. Wise	USA

LOCAL ORGANISING COMMITTEE

A. Araman	Istanbul University
B. Arıca	Hacettepe University
S. Çalış	Hacettepe University
A. A. Hıncal	Hacettepe University
H. S. Kaş	Hacettepe University
F. Öner	Hacettepe University
L. Öner	Hacettepe University

UNDER THE AUSPICES OF

- TÜBİTAK—The Scientific and Technical Research Council of Turkey

- Marmara University

- Hacettepe University Development Foundation

- Fako İlaçları
- Pharmacia Upjohn
- Planta Pharma

- Ali Raif İlaç San. A.Ş.
- Berksam İlaç Tic. A.Ş.
- Bilim İlaç Sanayii
- Karadeniz Ecza Deposu
- Organon İlaçları A.Ş.
- Roche İlaçları
- Schering–Plough Tıbbi Ürünler Tic. A.Ş.
- Wyeth İlaçları A.Ş.

FOREWORD

Advancing with Biomedical Engineering

Today, in most developed countries, modern hospitals have become centers of sophisticated health care delivery using advanced technological methods. These have come from the emergence of a new interdisciplinary field and profession, commonly referred to as "Biomedical Engineering." Although what is included in the field of biomedical engineering is quite clear, there are some disagreements about its definition. In its most comprehensive meaning, biomedical engineering is the application of the principles and methods of engineering and basic sciences to the understanding of the structure–function relationships in normal and pathological mammalian tissues, as well as the design and manufacture of products to maintain, restore, or improve tissue functions, thus assisting in the diagnosis and treatment of patients. In this very broad definition, the field of biomedical engineering now includes:

- System analysis (modeling, simulation, and control of the biological system)
- Biomedical instrumentation (detection, measurement, and monitoring of physiologic signals)
- Medical imaging (display of anatomic details or physiologic functions for diagnosis)
- Biomaterials (development of materials used in prostheses or in medical devices)
- Artificial organs (design and manufacture of devices for replacement or augmentation of tissues or organs)
- Rehabilitation (development of therapeutic and rehabilitation procedures and devices)
- Diagnostics (development of expert systems for diagnosis of diseases)
- Controlled drug delivery (development of systems for administration of drugs and other active agents in a controlled manner, preferably to the target area)
- Tissue engineering (use of functional, healthy cells from different sources, usually with extracellular components, either natural or synthetic, in the development of biological substitutes for restoration or replacement of tissue function)

With all of the exciting fields of application mentioned above, biomedical engineering is today providing the methods, materials, and devices for the effective and efficient solution of the problems in our present health care system by using existing engineering technology and systems methodology.

The number of health care facilities in Turkey is also rapidly growing and, in parallel to the advances in the developed countries, medical staff, especially in modern hospitals, are using products of the mainly imported sophisticated biomedical technology. These activities abroad have induced both Turkish scientists in the academia and the experts in the related industries to contribute to the methodology and techniques of this novel field of science and technology, or to develop new, local alternatives in our institutions.

In order to contribute to the activities in this rapidly growing field in Turkey, we initiated a series of symposia on "Biomedical Science and Technology" about four years ago. The fourth one, with international status, was held on September 15–17, 1997, in İstanbul. The main emphasis was on drug delivery and targeting, but other applications were also included. We believe and hope that, as previously, these meetings would bring the scientists and engineers from different disciplines, medical staff from clinical medicine, and experts from related industries together and trigger the development of this existing field of science and technology in Turkey.

Erhan Pişkin and Vasıf Hasırcı
Coordinators

PREFACE

This book contains the full papers of the invited lecturers and the oral presentations of the 4th International Symposium on Biomedical Science and Technology (BIOMED-4), held in İstanbul-Turkey between September 15–17, 1997. This symposium is organized with the intention of exchanging information with scientists all over the world in this growing interdisciplinary field of biomedical science and technology and of encouraging young researchers to find a place for themselves in this wide spectrum of interesting subjects. The symposium exposed to the participants several aspects of biomedical science and technology ranging from polymers, adjuvants, and vaccines to pharmaceutical, biotechnological, dental, and medical applications. The careful blend of the plenary lectures provided the participants with the opportunity to receive an overall view of these subjects. The main subjects covered at this symposium were application of microencapsulation, recent advances in nanoparticles, biodegradable polymers, pH-sensitive polymers, phospholipid-stabilized emulsions, drug carriers for biotechnology-derived products, tissue engineering of liver, adjuvants in vaccines, perfluorocarbons in wound healing, and polymers used in dentistry, orthopedics, and urology.

We believe that the presentations of the invited lecturers and oral presenters compiled in the following pages will contribute significantly to future studies applying biomedical science and technology.

The Organising Committee wishes to thank TÜBİTAK (The Scientific and Technical Research Council of Turkey), Marmara University president's office, Hacettepe University Development Foundation, and the national and international pharmaceutical companies that supported the activities of this symposium. The editors wish to thank Pharm. Betül Arıca (M.Sc) for her computer assistance, and would like to express their appreciation to Plenum Press for the opportunity to publish the proceedings and, in particular, to the authors of the papers, without whose contributions this work would not be available to those who are interested in this area.

We hope that this book will be helpful to those working in this interdisciplinary field.

A. Atilla Hıncal and H. Süheyla Kaş

CONTENTS

POLYMER BASED DRUG DELIVERY SYSTEMS

Erhan Pişkin

Chemical Engineering Department
Bioengineering Division
Hacettepe University
Beytepe, 06530 Ankara, Turkey

DEFINITIONS

Drug delivery systems are used to improve therapeutic efficacy and safety of drugs by delivering them at a rate dictated by the needs of the body over the period of treatment, and to the site of action, which may reduce size and number of doses, side-effects, and biological inactivation and/or elimination.[1-15]

As schematically illustrated in Figure 1, in conventional drug use (most commonly by oral or parenteral administrations) the drug concentration profile in plasma shows a series of peaks and valleys, depending on the number and amount of doses and the inactivation and/or elimination rate of the drug. Note that undesirable side effects may be associated with these fluctuations in drug level. By "sustained release dosage forms", it is possible to prolong therapeutic blood or tissue levels of the drug for an extended period of time (Figure 1). However, these forms are unable to control either the rate or site of action. While "controlled release systems" are capable of delivering drugs at predetermined rates and for predetermined period of time. Figure 1 shows that a constant drug concentration level in the blood plasma may be achieved by controlled release dosage forms which release the drug molecules at a constant rate (i.e. commonly known as zero-order kinetics). However, it should be noted that it is usually difficult to achieve constant drug levels due to unpredictable depletion rate of the drug in vivo. In order to use the benefits of localise drug action, drug delivery systems may be implanted adjacent to the target tissues or cells, or "intelligent carriers" which are capable to localise at the target area after intravascular injection may be tailor-made. The later case is commonly known as "drug targeting" or "site-specific drug delivery".

FACTORS EFFECTIVE ON DESIGN AND PERFORMANCE

Table 1 gives the important factors affecting both design and performance of drug delivery systems; i.e. drug properties, route of administration, target site and others, which are briefly discussed below.[1-15]

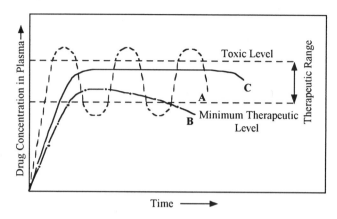

Figure 1. Plasma drug concentration profiles: Conventional dosage forms (A); Sustained (B) and Controlled Release Dosage Forms (C).

Drug Properties

Solubility (which depend on several parameters, such as chemical structure, molecular size, etc.) of a drug affects significantly both incorporation of the drug molecules within the carrier matrices (i.e. drug loading) during preparation, and their release and adsorption at the target area (i.e. bioavailabilities) during use. For instance, hydrophilic drugs can readily be dissolved in the hydrophilic drug carrier matrices and in the aqueous biological medium, however they cannot pass through the hydrophobic cell membrane easily at the target site. While hydrophobic drugs exhibit much lower solubility in the biological media, however, they can readily be dissolved or distributed in the hydrophobic drug carrier matrices, and can easily penetrate through the lipid layers of the cell membranes.

Therapeutic index for a drug, which is defined as follow, is generally used as a relative (not absolute) index to describe the margin of safety:

Therapeutic Index = Toxic Dose / Effective Dose

The release rate of the drug from the delivery system should be arranged in order to keep the drug concentration within the safe and effective limits, which are illustrated in Figure 1. Therapeutic indices of some selected drugs are given in Table 2.[2] Therapeutic indices, and the safe and effective levels of the drugs should be used together in the decision of margin of safety.

Table 1. Factors influencing design and performance
of drug delivery systems

• Drug properties
• Route of drug delivery
• Targeting
• Others: course of disease, patient, type of therapy, etc.

Table 2. Therapeutic indices of some selected drugs and their plasma safe and effective levels

Drug	Therapeutic Index	Effective Dose-Toxic Dose
Digitoxin	2.1	14-30 µg/L
Lidocaine	3.3	1.5-5 mg/L
Nortriptyline	2.8	50-140 µg/L
Procainamide	2.0	4-8 mg/L
Quinidine	2.5	2-5 mg/L
Salicylates	2.0	150-300 mg/L
Theophylline	2.0	10-20 mg/L

The biological half-life of a drug, which depends on its elimination, metabolism and distribution, plays one of the major roles in the design and use of drug delivery systems. Drugs may loose their activities during preparation of the delivery system, and especially in the biological media to which they are exposed. Note that, drug delivery formulations may offer an advantage in the increase of stability of the drugs in the biological media. Metabolism of a drug can either be inactivate the drug or convert an inactive drug to an active metabolite. Metabolic alteration of a drug can occur mainly in the liver. If the metabolism is predictable, this information may be used in the design of the drug delivery system and its in vivo performance. Distribution of drugs into tissues is usually an important factor in the overall drug elimination kinetics. Interaction of drugs with blood proteins may increase their duration of action. In an ideal system, the released drug molecules should completely be adsorbed by target tissue. In other words, the rate limiting step in drug delivery should be the release from the carrier system. However, this cannot be achieved in many cases due to degradation (i.e. elimination and metabolism) of the drug, protein-drug interactions, physical losses, etc.

Route of Administration

Drug delivery systems may be administrated via different routes as shown in Table 3, depending upon the aim of the therapy and the type of the delivery system (including the drug).[1-15] The route of administration has a pronounced impact on the therapeutic efficiency of the drug.

The oral administration of drugs is by far the most common route. Chemical degradation in the stomach, limited residence time in gastro-intestinal tract (GIT), metabolic degradation, especially in the liver (by the first-pass metabolism) are the main limitations of the oral administration. Sustained release formulations may eliminate chemical degradation in the stomach and prolongs of the GI transit time.

Parenteral formulations may be administrated in different routes. Drugs incorporated within nano- or micro beads or capsules, or soluble polymer chains or micelles may be injected intravenously. Targeting to the desired organ (or tissues) may be achieved in this

Table 3. Routes of administration

- Oral
- Parenteral: Intravenous, subcutaneous, intraperitoneal, intramuscular
- Transdermal
- Vaginal; cervical; intrauterine; rectal
- Buccal; sublingual; nasal; ocular; pulmonary

approach. Intramuscular or intraperitoneal injection, or injection or implantation into subcutaneous tissues of drug delivery systems have widely been used for local therapy.

Transdermal drug delivery systems, which include simple patches carrying the drug in a reservoir and containing a polymeric membrane for controlling of the drug release, are gaining attention recently. The low skin permeability of most drugs is the main limitation of the use of this route.

Vaginal, cervical and intrauterine drug delivery devices with different shapes have been mainly utilised for controlled release of hormones aimed at obtaining contraception for prolonged period with minimal side effects. The rectal route have been used for the drugs which are inconvenient for oral application, or are eliminated with the hepatic first pass metabolism. Due to mainly defecation, rectal route has been found limited number of uses.

Drugs can be adsorbed readily and rapidly through the mucosa in the oral (sublingually or bucally) and nasal cavities. Due to easy of administration and rapid adsorption, these routes have been considered as effective alternative drug delivery routes. The limited contact time is the main disadvantage of these routes. Ocular drug delivery systems have been used especially for the drugs with very short biological half-lives to optimise precorneal delivery. Drug delivery of aerosolised drug through respiratory system has been studied mainly for local treatment of inflammation and cancer in the lungs. However, the low efficiency of the delivery have been pointed out.

Targeting

Site-specific drug delivery, commonly known as "drug targeting", is one of the main aims in the applications of drug delivery systems, which may be classified different ways as given in Table 4. The most common method of classification is according to the mode of application, i.e. passive or active targeting. In this approach, drugs are loaded within the polymeric nano-or microcapsules or spheres, or attached to the macromolecules (e.g. antibodies, DNA) or soluble polymers (e.g. poly(α-aminoacids) or their micelles (e.g. poly(lactic acid)-poly(ethylene glycol) copolymers). These carriers are injected intravascularly and are therefore, targeted to the sites, passively or actively. In passive targeting, carriers from 0.01 μm to 100 μm in diameter are distributed in the vascular system and are captured in the specific capillary beds according to their sizes. For instance, if carriers larger than 20 μm are injected intravenously they are captured in the capillaries of lungs, therefore, by this way one could target these polymeric carriers to the lungs for the selected treatment of the disease there. In the active transport, commonly, ligands (e.g. antibodies, antigens, receptors) with specific recognition abilities are incorporated on the carrier surfaces, which leads targeting more specifically to the selected sites.

Table 4. Classifications of site-specific drug delivery

- Levels (i.e. first, second or third)
- Orders (i.e. organ, part of an organ or intracellular)
- Generation (i.e. organ, tissue or cellular)
- Mode (i.e. passive or active)

Others

In addition to the factors discussed above, the state of disease (i.e. changes during the course of application), the patient (whether the patient is ambulatory or bedridden, young or old, etc.), and the type of therapy (acute or chronic, time course, etc.) are the factors one should carefully take into consideration for the design and performance of the aimed drug delivery system.

FORMS

Both nondegradable and biodegradable polymers are used as carriers in different forms in drug delivery systems as given in Table 5. Injectable polymeric drug delivery systems are prepared in the form of nano- or microcapsules or spheres, or soluble or micelle forming polymer formulations. Membranes (i.e. hollow fibers, sheets, tubings or microcapsules); single or multiflament fibers, and wowen or non-wowen, or knitted fabrics or moulded objects with a variety of shapes (e.g. disks, needles, buttons, hemispheres, etc.) are used as implantable drug delivery systems.

Polymer based drug delivery systems may be injected into the cardiovascular blood system (the intravascular system), or may be injected or implanted into the tissues (the extravascular system). An ideal polymeric carrier system for use in intravascular system is expected: (i) to be blood-compatible (do not cause undesirable events, e.g. thrombus or emboli formation, complement activation;[16] (ii) to circulate in the blood stream without causing embolization at capillaries; (iii) to escape from excretion in the kidneys; (iv) not to be uptaken by the reticuloendothelial system nonselectively; (v) to release the drug, preferentially at the target area at a desired rate; and (vi) to degrade in vivo during or after drug release.

When a polymer based drug delivery system is injected or implanted extravascularly, the most important consideration is its tissue compatibility.[17] Foreign objects in tissues always generate a response. The major tissue response in the extravascular system is the inflammatory process. Inflammation may be induced biologically, chemically or physically.

Many proteins and cells are involved in this very complex process. The chemical characteristics of the foreign object, or in the long-term, released substances and/or the biodegradation products, may be responsible for foreign body reactions. Cell ingestion, fibrous encapsulation or fibrous ingrowth may occur depending on the geometry, configuration and size of the drug delivery system. Severe inflammation or excessive

Table 5. Forms of polymer based drug delivery systems

Injectables

- Nano- or microcapsules or spheres
- Water soluble polymers (or macromolecules)
- Micelle forming polymers

Implantables

- Membranes: Hollow fibers, sheets, tubings or microcapsules
- Fibers and fabrics: Single or multiflament fibers, woven or non-wowen or knitted fabrics
- Moulded objects: Disks, needles, buttons, hemispheres, or other shapes

---

fibrosis may even cause tissue necrosis, granulomas or tumorgenesis. Besides these very undesirable effects on the host tissue, these phenomena may lead the lost of the function of the delivery system, i.e. controlled release of the drug.

SYSTEMS AND RELEASE KINETICS

In polymer based drug delivery systems, the drug molecules are dispersed or dissolved, encapsulated or bound in or on a polymeric carrier.[1-15] Figure 2 shows release of drug from a typical "diffusion controlled matrix system", also commonly known as a monolithic device, in which drug molecules are dissolved or dispersed within a polymeric matrix. As seen in this figure, release rates from these type of systems are usually not constant (diminishe with time). Both polymer properties (e.g. crystallinity, T_g, cross-linking, porosity) and polymer-drug interactions (e.g. solubility, compatibility) may be effective on the release rate.

In the "diffusion controlled membrane systems", in which drug saturated solutions (usually containing extra solid drug) or suspensions are entrapped within polymeric membranes (in the form of microcapsules or hollow fibers), and constant release rates may easily be achieved (Figure 2). Time-lag or burst effect is observed at the beginning, and a reduction in the release rate with time is observed at the last part of the release. The drug release rate may be controlled by arranging properties of the polymeric membrane.

Both membrane and matrix systems may be biodegradable or swellable, in which polymer matrix is degraded or swell, respectively (Figure 2). The drug release in these cases is more complex. In addition to diffusion, the drug release rate may be controlled by degradation rate or swelling rate and extent of the polymeric matrix (or membrane), or both.

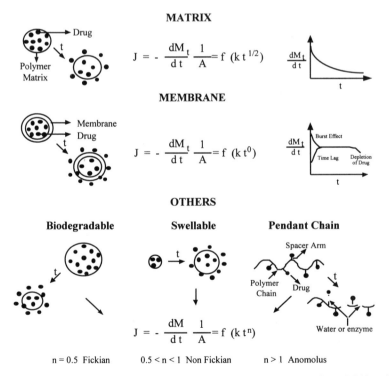

Figure 2. Polymer based drug delivery systems and release kinetics. Here: J: flux; dM$_t$/dt: release rate; A: surface area; t: time, k: rate constant; n: power and f: function.

Drugs may be incorporated with water soluble or micelle forming natural and synthetic, biodegradable or nondegradable (with a degradable pendant chain) macromolecules or polymers (Figure 2) are also used to prepare polymer based drug delivery systems, especially for targeting.

CRITERIA FOR SELECTION OF POLYMERS

Polymers must meet certain criteria and regulatory requirements before they can be qualified for use in drug delivery applications. First of all they should be readily purified, fabricated and sterilised easily by conventional methods. The main disadvantage of the polymers is the extractables in their structures which may come from the production of the polymers or their processing into the final form. Extractables, such as initiators, stabilisers, emulsifiers, unreacted monomers or oligomers, and other additives (e.g. plasticizers, fragments of fillers, dyes) may leach out during application, and may cause important side-effects.

Polymers should exhibit the biomechanical properties (in tension, compression and shear) necessary for the specific application. Therefore, they should have desirable physical structures (e.g. crystallinity, entanglement, equilibrium swelling), and other desirable properties such as permeability, elasticity, etc.

Polymers must be biocompatible. In other terms, they and their degradation products (if they are biodegradable) must not induce undesirable host reactions (e.g. thrombosis, inflammatory reactions, tissue necrosis, toxicity, allergenic reactions, carcinogenesis). All these properties are dependent upon time, temperature and environmental conditions, and may change within the physiological environment (in vivo) during application of the system. A drug delivery system must be safe and efficient during the course of delivery.

POLYMERS

Both nondegradable and biodegradable polymers are used in drug delivery systems.[1-21] Here, some examples of nondegradable and biodegradable polymers are given.

Nondegradable Polymers

Both hydrophilic and hydrophobic polymers have been utilised in drug delivery systems (Table 6). Hydrogels are cross-linked three-dimensional structures of, mainly, water soluble polymers. They do not dissolve in aqueous media, but can absorb water more than 30% of their weights, and therefore do swell to a certain extent depending upon the degree of cross-linking. When hydrogels are used as carrier matrices in dry form, they swell during the course of drug release,therefore, not only the diffusion but also swelling rate affects drug release kinetics, as mention before. The main advantage of hydrogel matrices is their very high biocompatibilities. Mainly hydrophilic, but also hydrophobic drugs can be incorporated with these polymer matrices. Polyhydroxyethylmetharylate (PHEMA) and other hydrogels have been extensively used in medical applications including drug delivery systems.

Silicones (especially Medical Grade Silastic™ by Dow-Corning) are one of the earliest and have been one of the most successful polymers in medical applications. They are the major amorphous soft (rubbery) polymers used for controlled release of hydrophobic drugs. With high stabilities, biocompatibilities and inertness, ethylene-vinyl acetate copolymers have also been utilised in controlled release formulations. They are mainly amorphous polymers, but may have crystalline regions, which may further control the drug release rate. Polyurethane elastomers are block copolymers consist of soft segments of polyols (e.g.

Table 6. Major nondegradable polymers for drug delivery systems

Hydrophilic polymers:	Hydrophobic polymers:
Polyhyrdoxyethylmethacrylate	Silicones
Polyethylene oxide	Ethylene-vinyl acetate copolymers
Polyvinyl pyrrolidone	Polyurethane elastomers
Polyvinyl alcohol	
Polyacrylamide	

polyethylene oxide and hard segments of isocyanates (e.g. toluene diisocynate), which have also been investigated in drug delivery systems. Note that all these amorphous elastomeric polymers may be crosslinked to control the drug release.

Biodegradable Polymers

Drug delivery systems are also prepared from biodegradable polymers that are degraded in the body during the course of drug release or after the release completed. In many of the medical applications of polymeric biomaterials, biodegradation is an undesirable event, which causes alteration of the material properties. However, in the case of drug delivery, biodegradation is usually considered as a desirable property. Because many types of implantable or injectable drug delivery systems only function for a certain period of time in vivo. If a biodegradable matrix is used, accumulation of residual polymer is avoided, and hence, the need for a secondary surgery to remove the implant is eliminated.

The main mechanism of in vivo degradation of polymers used in drug delivery systems is "hydrolytic degradation" in which enzymes may also take role (i.e. "enzymatic degradation"). There are number of parameters which influence the hydrolytic degradation of polymers in vivo, as outlined in Table 7. Therefore, it is a highly complex phenomena.

Briefly, if the water molecules penetrates easily within a polymeric matrix (i.e. hydrophilic polymers), degradation occurs rapidly, and most probably, within the bulk (i.e. bulk erosion). Hydrophobic polymer matrices are degraded much slower than hydrophilic polymers, and from the surface (i.e. surface erosion). Chemical structure of the matrix has a pronounced effect on degradation. Hydrolytically unstable bonds (e.g. ester bonds) are degraded much faster. Polymers with smaller molecular weights are usually degraded faster than those with higher molecular weights. Morphology of the polymer is also important. Crystalline regions are usually degraded slower than amorphous regions, because water molecules cannot easily penetrate crystalline regions. Polymers which exhibit a glass transition temperature lower than the body temperature are rubbery in vivo conditions, degrade faster than the polymers which are glassy in vivo conditions, mainly due to flexibilities of the polymer chains in rubbery phase.

The residuals or additives (e.g. monomers, drugs) within the polymeric structure may change the course of degradation. Geometrical factors (e.g. size, shape, and surface to volume ratio) influence the degradation. High surface to volume ratio area means high water penetration, which leads high degradation rates. The degradation products (e.g. the carboxylic acid end groups generated during the hydrolysis of ester bonds) and enzymes do usually enhance the degradation rate, i.e. the so-called, autocatalytic and enzymatic degradation, respectively. Environmental factors, e.g. site of implantation or injection, and pH, ionic stregnth and temperature of the degradation medium have profound effects on both rate and extend of hydrolytic degradation of polymeric materials, as expected.

Table 7. Important factors effective on hydrolytic degradation of polymers

- Water permeability
- Chemical Structure
- Molecular weight
- Morphology
- Glass transition temperature
- Additives
- Device dimensions
- Mechanism of hydrolysis
- Environmental Factors:
  Site of implantation or injection;
  pH, ionic strength and temperature of the degradation medium

The major natural and synthetic biodegradable polymers used in drug delivery formulations are briefly introduced in Table 8.

Albumin. Albumin is the main blood plasma protein. Blood-compatibilities of blood-contacting biomaterials may significantly be improved by attaching albumin molecules on biomaterial surfaces. Because of its very high blood-compatibility it has been evaluated as a potential carrier matrix in intravascularly injectable drug delivery systems. Albumin microspheres are usually prepared by emulsion stabilization (i.e. emulsion cross-linking). Briefly, an aqueous mixture of albumin and drug is dispersed within a hydrophobic vegetable oil (e.g. olive oil) or in a relatively hydrophilic PMMA solution. Then, cross-linking is achieved by thermally (at 100-180°C) or chemically (by using chemical cross-linkers, e.g. formaldehyde, glutaraldehyde) through disulphide bonds or formation of lysine-alanine cross-links. Albumin particles in a wide range (from 1 mμ to 1000 mμ in diameter) can be produced by this method. Particle size is decreased by increasing the agitation rate and aqueous phase/organic phase ratio. Drug loading is usually below 10% (w/w). However, higher loadings (up to 20%) may be achieved for the drugs which interacts with albumin strongly. Drug release rate from these matrices decreases with the increase in cross-linking density, albumin-drug interaction, and particle size. Both hydrophobic and

Table 8. Major biodegradable polymeric biomaterials

Natural biodegradable polymers:

Albumin
Collagen & gelatine
Chitin & chitosan
Others: Fibrinogen, antibodies, dextran, alginate, casein, cellolose, strach

Synthetic biodegradable polymers:

Poly(a-hydroxy acids)
Poly(a-amino acids)
Poly(e-caprolactone)
Poly(ortho esters)
Poly(anhydrides)
Poly(alkyl 2-cyanoacrylates)

hydrophilic drugs, such as L-epinephrine, adrymaycin, 5-fluorouracil, insulin, epirubicin, prednisolon, have been loaded within albumin particles. Low loadings, thermal decomposition of drug during thermal cross-linking, interaction of drug with chemical cross-linkers, enzymatic degradation of the matrix, which leads bulk erosion (i.e. usually is an uncontrollable and unpredictable degradation) are noted as the main disadvantages of albumin carriers.

Collagen and Gelatine. Collagen is the most abundant protein in the animal kingdom, contributing about 30% of total body proteins in the vertebrate. It is present throughout the body, but especially in skin, tendon and bone. As the major component of the extracellular matrix, collagen occupies a key position in the molecular architecture of higher animals, providing strength and structural stability to various tissues. Scheme 1 gives the structure of collagen. Gelatine is the denatured form of this fibrous protein. Reconstituted collagen biomaterials within a number of forms, such as sheets, tubes, sponges, powders and fleeces have found diverse applications in medicine. Reconstituted collagen can be prepared by physical or chemical crosslinking (i.e. so-called "tanning") of tropocollagen. Short wave length UV irradiation (254 nm) or dehydrothermal crosslinking (in vacuum for several days at temperatures upto 100°C) are two physical crosslinking methods which have been used. Note that a lower degree of crosslinking is achieved by these physical methods.Trivalent metals such as aluminium or chromium are used in the tanning of collagen biomaterials (such as catgut surgical sutures) by forming ionic intra- and intermolecular bonds. Stability of this type of crosslinks is low.

Chemical crosslinking through covalent bonds can be achieved in a number of ways. Crosslinking of collagen with glutaraldeyde is the most extensively used method. Note that crosslinking stabilize the structure, control the rate of degradation (therefore the drug release from these matrices), improve the mechanical properties, and also suppress the antigenicity of collagen.

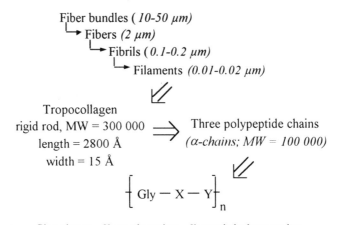

Fiber bundles (*10-50 μm)*
↳ Fibers *(2 μm)*
  ↳ Fibrils *(0.1-0.2 μm)*
    ↳ Filaments *(0.01-0.02 μm)*

Tropocollagen
rigid rod, MW = 300 000 ⟶ Three polypeptide chains
length = 2800 Å   *(α-chains; MW = 100 000)*
width = 15 Å

$$\left[Gly - X - Y \right]_n$$

Gly : glycine   X: mostly proline   Y: mostly hydroxy proline

Scheme 1. Structure of collagen.

Collagen and its denatured form gelatine have been used to prepare drug delivery systems, which are administrated subcutaneously, intramuscularly or intraperitoneally, and intravascularly with liposomes. However, they suffer from poor reproducability. They are biodegraded both hydrolytically and enzymatically (by collagenase, and other enzymes, e.g. elastase, cathepsin G, etc.), therefore it is usually difficult to predict the degration and drug release rate.

Chitin and Chitosan. Chitin is the major polysaccharide of the shells of crustaceans (Scheme 2). Chitosan is the deacetylated derivative of chitin. They are biodegradable, hemostatic, and promote the cell growth. They have been used in the applications similar to collagen. Cross-linked forms have also been evaluated in drug delivery systems.

Other Natural Polymers

Other polymers (or macromolecules) obtained from a variety of natural sources; such as fibrinogen, antibodies, dextran, alginate, casein, cellulose, starch, etc. have also been investigated as carrier matrices for drug delivery systems.

Poly(α-hydroxy) Acids. Copolymers and homopolymers of poly(α-hydroxy) acids form an important class of biodegradable polymers. Scheme 3 gives the most widely used members of this family. Polyglycolic acid (PGA) is a semi-crystalline polymer (crystallinity: up to 50% , T_g: 36°C and T_m: 225°C). In contrast to other members of the family, PGA is soluble only in few organic solvents (e.g., hexafluoroisopropanol and hexafluoroacetonesesquilhydrate). Its limited solubility and high melting temperature are considered as important limitation for processing.

Lactic acid is a chiral molecule, it exist in two stereoisomeric forms, L and D, with similar intrinsic chemical properties, but opposite configurational structures, therefore they are optically active. DL-lactic is a racemic mixture (an equal mixture) of D- and L-lactic acid, therefore is optically inactive. In biomedical applications mostly poly(L-lactide) and poly(DL-lactide) have been studied. Poly(L-lactide) is a semi-crystalline polymer (crystallinity: up to 40% , T_g: 57-60°C and T_m: 184°C) while poly(DL-lactide) is fully amorphous and therefore has only a T_g value of 54-59°C. Poly(DL-lactide) is soluble in most common organic solvents such as THF, acetone, chloroform, benzene, while poly(L-lactide) can be dissolved mainly in chloroform and methylene chloride. Glycolide and lactides can also be copolymerized. Owing to difference in reactivities of these two dimers, copolymers having broad composition ranges can be produced. Figure 3 shows the effect of backbone structure of PLA/PGA copolymers on crystallinity.

Glycolic acid and lactic acid can be polymerized directly into homo or copolymers with low molecular weights by heating. Polyglycolides or polylactides or their copolymers with higher molecular weights, therefore with reliable mechanical properties can be synthesized from glycolides or lactides (i.e. cyclic dimers of glycolic acid and lactic acid, respectively) by ring opening polymerization.

Biodegradation of poly(α-hydroxy) acids are degraded in vivo mainly by hydrolytical degradation,which starts first in amorphous regions then continuous in crystalline regions.

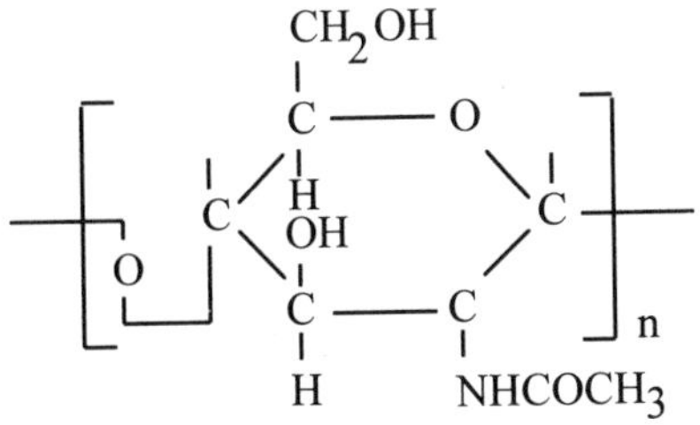

Scheme 2. Structure of chitin.

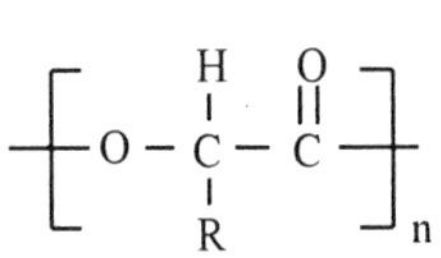

- Polyglycolide (R:H)
- Polylactides (R:CH_3)
  - → Poly(L-lactide) ⎤
  - → Poly(D-lactide) ⎦ → *(optical isomers)*
  - → Poly(D,L-lactide) → *(a rasemic mixture of L-, and D-)*
- Poly(lactide-co-glycolide)s

Scheme 3. Poly(α-hydroxy) acids.

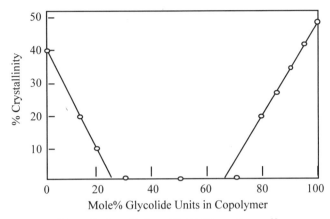

Figure 3. Crystallinity of PLA/PGA copolymers.[19]

They do degrade in vivo from weeks to years depending mainly on initial molecular weight, morphology (amorphous/crystalline phases), shape, and also backbone structure. Note that the carboxylic acid groups forming during degradation may autocatalyze the hydrolysis.

Poly(α-hydroxy) acids are toxicologically safe and commercially available. For example, the final degradation product of poly(L-lactide), i.e., L-lactic acid is a normal intermediate of carbohydrate metabolism in man. L-lactic acid is recycled by conversion to glycogen in the liver, while glycolic acid and D-lactic acid most probably excreted through the kidneys unchanged. Therefore, they have attracted a great attention as a biodegradable biomaterials. Polyglycolic acid and its copolymers with polylactic acid have been used for manufacturing of biodegradable sutures for many years. Clips, screws, pins, stables, meshes, and other fracture fixation implants have been manufactured from polylactic acids. PDLLA and other homo and copolymers of the poly(α-hydroxy) acid family have been tested in designing drug-releasing implants (with different shapes) and injectable microcarriers (i.e. spheres and capsules) as drug delivery systems. A wide variety of drugs exemplified in Table 9 have been incorporated in these matrices.

Poly(α-amino) acids. The chemical formula of two members of this polymer family is given in Scheme 4. These polymers can be manufactured by polymerization of the respective monomers. The ω-carboxylic acid groups, which are available for covalent attachment of active materials (e.g. drugs), are protected in the form of esters (e.g. methyl-benzyl) during polymerization.

Subcuteneaous delivery systems made of these polymers carrying several drugs (e.g. norgestrel, progesterone) have been studied. However the most interesting approach is their use to prepare prodrugs. In this application, drugs are covalently attached to poly(α-amino)

Table 9. Drugs incorporated within poly(a-hydroxy) acid matrices

- Antibiotics: e.g. ampicilin
- Antimalarials: e.g. sulpadiazine
- Chemotherapeutic agents: e.g. doxorubicin
- Contraceptive steroids: e.g. progesterone
- Local anaesthetics: e.g. dibucain
- Narcotic antagonists: e.g. naltrexone
- Other steroids: e.g. triamcinolone

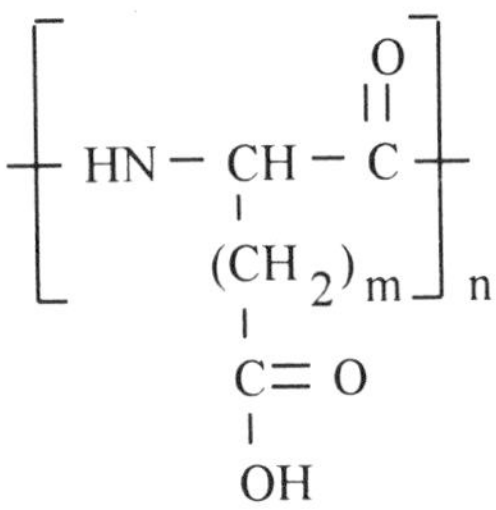

m=1: aspartic acid; m=2: glutamic acid

Scheme 4. Poly(α-amino) acids.

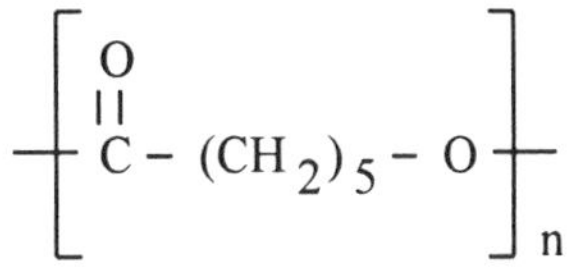

by ring-opening polymerization of a lactone
seven-membered ring

Scheme 5. Poly(ε-caprolactone).

acids through ω-carboxylic acid groups (directly or through a spacer arm) (see also Figure 2). These conjugates may then be injected into intravascular system, in which they are endocytosed by the cells (preferentially by the tumor cells) due to their macromolecular size. Antibodies (againts the membrane receptors of the target cells) may also be conjugated to the polymeric chain for more specific ("active") targeting.

Poly(ε-caprolactone). Polycaprolactone is another biodegradable polyester synthesized by ring-opening polymerization of a lactone (i.e. e-caprolactone) (Scheme 5). It is a hydrophobic, semi-crystalline polymer (T_g: -60°C and T_m: 63°C). It is in a rubbery state at room temperature, which contributes to its very high permeability for many therapeutic drugs. High loading and diffusion rates for hydrophobic drugs can be achieved in polycaprolactone matrices. Its in vivo degradation is much slower than poly(α-hydroxy acids), which starts nonenzymatic hydrolysis in extracellular matrix, and then contiuous enzymatically in the macrophages after phagocytosis. By copolymerization of polycaprolactone with poly(DL-lactide) gives flexible matrices with lower crystallinity and therefore higher degradation rates and higher drug releas, rates are achieved. Capronor™ is a typical 1 year implantable contraceptive device prepared from polycaprolactone.

Poly(ortho esters). Poly(ortho esters) have also been investigated for drug delivery applications. One of the important poly(ortho esters) used for drug delivery is known under the trade name of Chronomer™ or Alzamer™ (is used for controlled release of contraceptives), which is produced by condensation of 2,2-diethoxytetrahydrofuran and a dialcohol (Scheme 6). These polymers degrade by surface erosion, mainly due to the hydrophobicity of the matrix and acidic by-product that autocatalyse the degradation, and the additives used. Since surface erosion, zero-order release kinetics may be achieved by slab-like devices. These polymers have an advantage in formulating controlled release formulations of heat and solvent sensitive drugs such as hormones and growth factors. Simply, prepolymers with low molecular weights (therefore viscous liquid) are mixed with the drug and the drug delivery matrix is then solidified at a temperature of about 40°C. Low drug loadings and release rates may be considered as disadvantages.

Polyanhydride. Polyanhydrides are among the most hydrolytically unstable polymers used in drug delivery systems. A typical polyanhydride of bis-p-(carboxyphenoxy) propane (CPP) and sebacic acid (SA) is given in Scheme 7. This polymer have been tested in formulation of an implantable delivery system for the controlled release of a chemotherapeutic agent (i.e. BCNU) in the treatment of malignant brain tumours. Biodegradation rate of this polymer is much faster than poly(ortho esters), and occurs by surface erosion due to the hydrophobicity of the matrix and the fast degradation rates.

13

Poly(alkyl 2-cyanoacrylates). These polymers (Scheme 8) are produced by very rapid anionic polymerization at room temperature, in which hydroxyl ions initiate the reactions. They degrade very fast (few hours) in alkaline pH, while they are quite stable in acidic pH (usually around pH: 1-2). Lower homologous (e.g. methyl or ethyl) of poly(alkyl 2-cyanoacrylates) are degraded much faster than higher homologous (e.g. hexyl). It is generally accepted that due to very fast degradation rates (which means very fast accumulation of the degradation products, i.e. oligomers and formaldehyde at the surrounding tissue) lower homologs may lead toxicity. These polymers have been used as tissue adhesive and hemostatic agent for a long time. They have also been evaluated as carriers in the forms of nano- or micro-particles in intravascular drug delivery formulations.

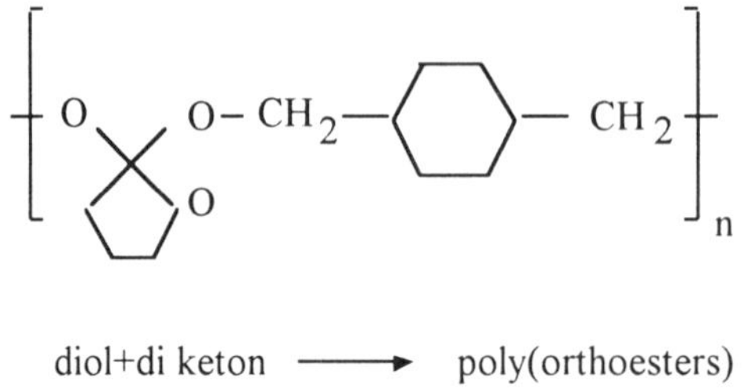

diol+di keton $\longrightarrow$ poly(orthoesters)

Scheme 6. Poly(ortho esters) (Alzamer™).

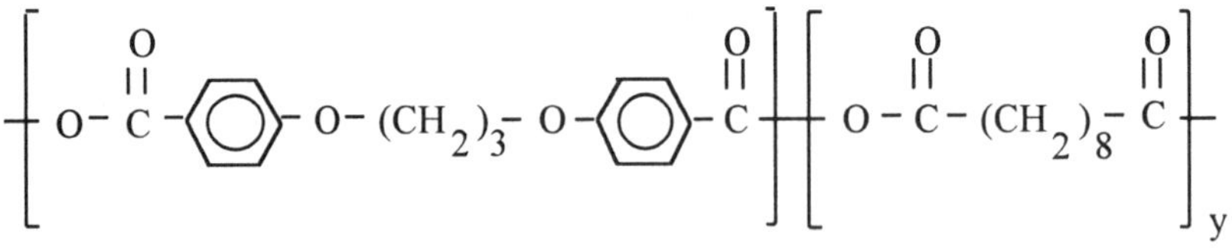

bis-p-(carboxyphenoxy) propane   sebacic acid

Scheme 7. Polyanhydride of CPP and SA.

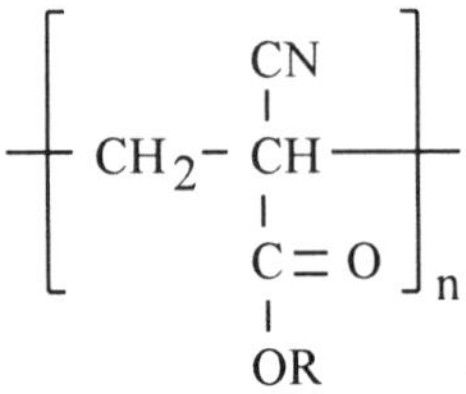

R: Alkyl group (e.g., methyl, hexyl)

n ~3000

Scheme 8. Poly(alkyl 2-cyanoacrylates).

Emerging Trends

Many of the current polymers and processing techniques need to be improved in order to produce polymers with better performances in biological media. An important trend in the related research and development is the synthesis of novel polymers, which would exhibit improved biocompatibility, and would be bioresponsive (Table 10).[9, 22, 23]

For improved biocompatibility the phenomena at the polymer-living system interface and the effects of polymer surface properties on these phenomena should be better

Table 10. Emerging trends

Polymers with improved biocompatibility:

- Studies for better understanding of the phenomena at the polymer-living system interface
- Development of new analytical methods, imaging and assaying techniques
  to study the interfacial phenomena
- Biological modification of polymer surfaces

Bioresponsive and biofunctional polymers:

- Preparation of polymeric matrices which would respond
  physical, chemical or biological stimuli
- Polymeric carriers with ligands which would have specific biological recognition abilities

---

understood. If the specific "needs" for improved polymer surface characteristics can be defined, then these surfaces may be tailor-made. For better understanding of the interfacial phenomena, new analytical methods, and imaging and assaying techniques are needed for analysing the polymer surfaces at molecular level (both at static and dynamic phases) and observe the behaviour of biological entities (e.g., proteins, cells) both on the surfaces and in the bulk.

Biological modification of polymer surfaces is one of the new trends in this field. Chemical and physical modification of the polymeric surfaces may significantly increase their biocompatibilities. Biological modification may be considered as one further step to improve their biocompatibilities, and also to add a biofunctionality to the respective surfaces. The main goal in this approach is to incorporate biological entities (e.g. heparin and heparin fragments, heparinase, urikinase, streptokinase, albumin and endothelial cells) onto polymeric material surfaces to create polymer-living system interfaces which are close to mother nature.

Another important trend emerging is to prepare stimuli-responsive polymers. These polymers may synthesised with the ability to respond to physical, chemical or biological signals (e.g. temperature or pH changes, electrical or optical signals, metabolites) in the biological environment. Therefore, release characteristics of these responsive systems may be better controlled and therefore, drugs may be delivered at predictable release rates in vivo.

A number of biological materials (e.g. enzymes, antibodies, and other proteins, antigens, amino acids, peptide sequences, nucleic acids, DNA and RNA sequences) may also be incorporated with polymers in order to prepare biologically active composites for site-specific drug delivery.

All these emerging trends are at a preliminary stage. However, with the exiting and stimulating results already achieved, it is believed that they will continue to be the most attractive directions in research and development of novel drug delivery systems with improved in vivo performance.

REFERENCES

1. A.C. Tanguary and R.E. Lacey (eds.), *Controlled Release of Biologically Active Agents*, Plenum Press, New York, 1974.

2. J.R.Robinson and V.H.L. Lee (eds.), *Controlled Drug Delivery-Fundamentals and Applications,* Marcel Dekker Inc., New York, 1978.

3. R.J. Kostelnik (ed.) *Polymeric Delivery Systems,* Gordon and Breach, New York, 1978.

4. E.J. Ariens (ed.), *Drug Design,* Vol.10, Academic Press, New York, 1980.

5. A.F. Kydonieus (ed.), *Controlled Release Technologies: Methods, theory and Applications,* Vols.I and II, CRC Press, Inc., Boca Raton, Florida, 1980.

6. Y.W. Chien (ed.), *Novel Drug Delivery Systems,* Marcel Dekker Inc., New York, 1982.

7. S.D. Bruck (ed.), *Controlled Drug Delivery,* Vols.I and II, CRC Press, Inc., Boca Raton, Florida, 1984.

8. R.Langer and D.L. Wise (eds.), *Medical Applications of Controlled Release,* Vols.I and II, CRC Press, Inc., Boca Raton, Florida, 1984.

9. J.M.Anderson and S.W. Kim (eds.), *Recent Advances in Drug Delivery Systems*, Plenum Press, New York, 1984.

10. S.S. Davis, L.Illum, J.G. McVie and E. Tomlinson (eds.), *Microspheres and Drug Therapy,* Elsevier Sci. Publ., Amsterdam, 1985.

11. P.Burri and A.Gumma (eds.), *Drug Targeting,* Elsevier Sci. Publ., Amsterdam, 1985.

12. E.Pişkin and A.S. Hoffman, eds., *Polymeric Biomaterials*, Martinus Nijhoff Publ., Dordrecht, 1986.

13. R. Baker (ed.), *Controlled Release of Biologically Active Agents*, John Wiley & Sons, New York, 1987.

14. L.Illum and S.S. Davis (eds.), *Polymers in Controlled Drug Delivery*, Wright Bristol, 1987.

15. F.H. Roerdink and A.M. Kroon (eds.), *Drug Carrier Systems*, John Wiley & Sons, New York, 1989.

16. D.F.Williams and D.J. Lyman (eds.), *Blood Compatibility* , CRC Press, Inc., Boca Raton, Florida, 1982.

17. D.F.Williams (ed.), *Fundamental Aspects of Biocompatibility*, Vols.I and II, CRC Press, Inc., Boca Raton, Florida, 1981.

18. M.Chasin and R. Langer (eds.), *Biodegradable Polymers as Drug Delivery Systems*, Marcel Dekker Inc., New York, 1990.

19. T.H.Barrows, Synthetic Bioabsorbable Polymers, In: "*High Performance Biomaterials*" M.Szycher, ed., Chapter 39, Technomic Publ.Comp., 1991.

20. D.Byron, *Novel Materials from Biological Sources,* Stockton Press, New York, 1991.

21. E.Pişkin and S.L. Cooper (eds.), *Biodegradation and Biodegradable Polymers*, VSP, Int.Sci. Publ., Zeist, The Netherlands, 1994 (in press).

22. D.D. Breimer and M.Midah, *Topics in Pharmaceutical Sciences,* Amsterdam Medical Press, Amsterdam, Holland, 1987.

23. E.Pişkin, *Biologically Modified Polymeric Biomaterial Surfaces,* Elsevier Sci. Publ., Essex, 1992.

RECENT ADVANCES AND INDUSTRIAL APPLICATIONS OF MICROENCAPSULATION

Simon Benita

Department of Pharmaceutics
School of Pharmacy
The Hebrew University of Jerusalem
P.O.B. 12065
Jerusalem 91120
ISRAEL

INTRODUCTION

Peptides and recombinant proteins of prime pharmacological importance with promising therapeutic applications are currently being produced on industrial scales owing to the marked progressed and major developments accomplished by the biotechnology industry. However the clinical efficacy of these promising sensitive potent molecules have been offset by their poor ability to penetrate the gastrointestinal barrier and their easy degradation by proteolytic enzymes limiting their administration only to the parenteral route. In addition to these constraints some of the peptides and most of the proteins are sensitive to heat, and organic solvent denaturation and exhibit very short biological half lives thus presenting a real challenge to the pharmaceutical scientist interested in designing injectable controlled release formulations of such fragile potent drugs. Most controlled release systems currently under investigations for polypeptides and recombinant proteins are injectable implantable devices comprised of protein microparticles within a polymeric matrix or mainly microcapsules of the matrix type.[1]

A large number of microencapsulation processes based mainly on coacervation, interfacial polymerization and spray-drying techniques have been extensively reported in the literature and well-revised in numerous textbooks.[2-4] These microencapsulation technologies, which generally require the use of organic solvents, have led to the development of effective products that are well accepted in the pharmaceutical market. Most of the successful commercial products have undergone careful testing for residual organic solvents in order to prevent any possible harmful effects being induced in the patient or user.

In this chapter emphasis is focused on the description of selected, modified processes leading to peptide or recombinant protein microparticle formation generally from coating polymers and copolymers of lactic and glycolic acids (PLA, PLGA) which are the sole biodegradable injectable excipients presently approved by international health authorities. The processes commonly used to produce PLGA microcapsules, which were developed to encapsulate small molecules, peptides and occasionally proteins, include phase separations, solvent evaporation and spray-drying. These processes, however employ elevated temperatures, surfactants or aqueous/organic solvent interfaces -- conditions that might denature most proteins. Thus modification of these methods has been recently proposed to encapsulate proteins without altering their integrity.

In addition to providing improved patient compliance and convenience, microparticulate controlled-release delivery systems can provide greater safety and efficacy as shown with the pioneer marketed long-acting, injectable microcapsule (matrix type) delivery systems of triptoreline (LHRH analogue), Decapeptyl®.[5] The success of the latter product has contributed to the considerable research conducted in the area of peptide and protein delivery using microencapsulation techniques.

Furthermore, the governmental authorities are imposing severe regulations to avoid organic solvent environment contamination during product processing, resulting in an increase in cost manufacturing. Due to these restrictive directives, scientists are now concentrating their efforts on the development of novel solvent-free microencapsulation technologies such as methods using fluids under supercritical conditions. These methods are more adaptable to the microencapsulation of potent active molecules, such as peptides, and proteins which are sensitive to heat, hydrolysis and pH. Research emphasis will probably focus in the near future on these promising techniques. Updated studies and achievements of these different microencapsulation techniques will be presented and discussed.

ORGANIC SOLVENT BASED TECHNOLOGY

Phase-Separation Methods

Polymer phase-separation, in non aqueous media, by nonsolvents or polymer addition, also referred to as coacervation, is an excellent technique for the entrapment of water-soluble drugs such as peptides or proteins into a final microcapsule product. Under specific experimental conditions, it was shown that the coacervation of PLGA dissolved in methylene chloride by the addition of silicone oil (phase separation inducer) resulted in the formation of microcapsules (matrix type) containing eventually a peptide suspended in methylene chloride at the beginning of the process.[6] The manufacturing process of the phase-separation coacervation technique is schematically described in Figure 1. The technique for phase separation has been used for the preparation of at least two marketed peptidic microencapsulated products. Triptoreline, a LHRH analogue, as previously mentioned (the coating polymer of which consists of PLGA) and used in the treatment of prostate cancer and endometrosis[5] and lanreotide (somatostatine analogue) approved for the treatment of acromegally.[7] Formulation parameters have been adjusted to allow a constant release in vivo over prolonged periods of time. Compared to spray-drying or solvent evaporation methods, this phase separation coacervation technique may protect the peptides and proteins from being altered by exposure to heat or from their partitioning out into the dispersing phases. However, residual solvent concentrations in the microcapsules, can be very high.[8,9]

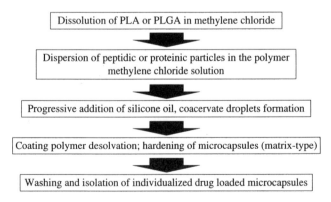

Figure 1. Schematic description of the microcapsule manufacturing process based on phase separation-coacervation technique.

Spray-Drying Technique

The principle of spray drying by nebulization rests on the atomization of a solution (containing the active ingredient and coating polymer to be dried) by compressed air or nitrogen through a dessicating chamber and drying across a current (or stream) of warm air. A schematic description of the manufacturing process using spray drying is presented in Figure 2. This technique has been widely used by the industry to protect sensitive substance from oxidations (e.g. fish oil, essential oils, vitamins).[10] The most widely used coating materials for protective purposes particularly in the case of volatile substances are maltodextrine and arabic gum.

Depending on the initial experimental conditions used, spray drying technique can lead to final microcapsules of the "reservoir type" or "matrix type" (i.e. widely defined in the literature as microspheres). If in the initial step of processing a solution or a suspension is nebulized in the form of an aerosol, the final product leads to microspheres (drug dissolved or dispersed in a polymeric dense matrix network) while the nebulization of an emulsion in the form of an aerosol results in the formation of microcapsules (drug dissolved in liquid core surrounded by a distinct polymeric envelope). To date, only bromocryptine (Parlodel® LAR) is commercially available in the form of injectable microcapsules (matrix type) of PLGA branched to D-glucose.[11] Other authors have made use of this process for the encapsulation of soluble proteins and peptides such as busereline, protireline using mixtures of PLGA and polyhydroxybutyric acid (PHB).[12] Thus, the use of spray drying for the formation of biodegradable microcapsules based on PLA or PLGA of various molecular weights containing low sensitive but potent peptidic drug amounts is a recent feature which merits further testing especially for thermolabile compounds.

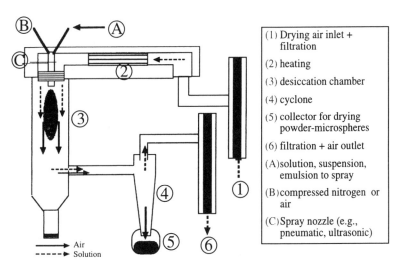

Figure 2. Schematic illustration of a laboratory spray dryer (4).

Solvent Evaporation Methods

The solvent evaporation technique is based on the evaporation of the internal phase of an emulsion under constant stirring. The active substance to be encapsulated is either dissolved or dispersed in the internal emulsion phase together with the coating polymer which is not soluble in the external phase of the emulsion. In the case in which we are particularly interested, i.e. microencapsulation of proteinic and peptidic compounds usually coated by DL-PLA or/and DL-PLGA polymers, an oil-in-water (O/W) emulsion is prepared. Usually the oil phase is an organic solvent (methylene chloride or ethyl acetate). Once the emulsion is stabilized by emulsifying agents (surfactants), agitation is maintained and the organic solvent evaporates after diffusion through the external continuous phase of the emulsion. This results in the formation of

19

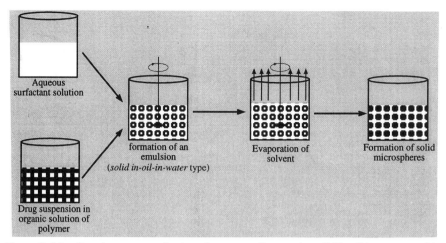

Figure 3. Manufacturing process of microspheres according to the solid-in-oil-in-water emulsion technique (S/O/W).

solid microcapsules of the matrix-type in which the drug is either dispersed or dissolved in the polymeric matrix network. Thus, DL-PLA and DL-PLGA microcapsules were first commonly prepared by this organic solvent evaporation process.[13,14] However, this technique has poor encapsulation efficiency of the water-soluble core materials. Many peptides and proteins are water-soluble and poorly soluble in organic solvents. Encapsulation efficiency of these water-soluble compounds within DL-PLGA and DL-PLA microcapsules has been improved by using organic phase separation[15,16]

Thus it can be deduced that the o/w emulsion technique is mostly adapted to the microcapsulation of lipophilic active ingredients unless the aqueous external phase is modified by addition of specific salts to decrease the polypeptide aqueous solubility as recently done for the microencapsulation of rismorelin porcine, a potent analog of growth-hormone releasing hormone.[17] The peptide or protein particles are then dispersed in the organic phase prior to emulsification as shown in Figure 3 resulting in the formation of a solid dispersed in the oily internal phase of an oil-in-water emulsion (S/O/W). The evaporation of the organic solvent leads to the formation of microspheres (matrix-type microcapsules) as shown in Figure 3. A similar approach was used by Cleland and Jones to encapsulate Interferon-γ.[18] The major objective of these authors was to develop formulations that stabilize recombinant proteins in organic solvents commonly used for microencapsulation, methylene chloride and ethyl acetate. They found that even drug protein powders must contain a stabilizing excipient to prevent denaturation during microencapsulation. Stability of the drug protein in organic solvents was increased through the addition of hydrating excipients capable of forming hydrogen bonding with the proteins.[18] However, Ogawa and coll.[19] proposed another approach for the efficient encapsulation of water-soluble active principles by the solvent evaporation technique in aqueous continuous phases through the formation of a multiple emulsion of the type water-in-oil-in-water (W/O/W). Their method was adopted for microencapsulation of water-soluble drugs[19,20,21] and vaccines.[22,23] The peptidic drugs are dissolved in the primary aqueous phase of the multiple emulsion as schematically shown in Figure 4. The organic phase acts as a barrier between the two aqueous compartments preventing the diffusion of the proteins and peptides toward the external aqueous phase. Evaporation of the organic solvent leads to the formation of the solid drug loaded microcapsules (microspheres) as illustrated in Figure 4. This technique with appropriate modifications and improvements has been successfully used for the encapsulation of a peptide analogue of LHRH, leuprolide acetate (marketed under the name of Enantone®). The biodegradable microcapsules were shown to release progressively the peptide over a 3 month period.[24] The W/O/W double emulsion technique was recently used to encapsulate proteins such as insulin,[25] bovine serum albumine[26] and rh-erythropoietin.[27] It appears that under adequate experimental conditions and addition of various protein stabilizing excipients degradation and denaturation of the proteins can be minimized during the microencapsulation process using the W/O/W double emulsion technique. More recently insulin was also

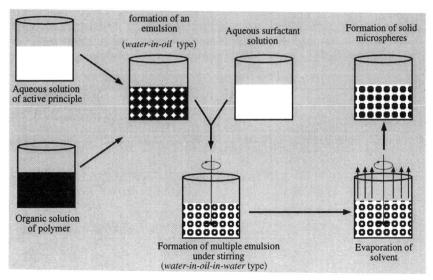

Figure 4. Manufacturing process of microspheres according to the multiple emulsion technique (water/oil/water).

incorporated in poly (DL-Lactide) microcapsules and the process was optimized to allow an adequate *in vivo* prolonged release profile following parenteral administration.[25]

Thus several methods of microencapsulation have been described in the literature. The choice of a method depends on the physical and chemical properties of the drug and polymer. However, a feature shared by most of the methods is the use of organic solvent and high shear. Exposures to organic solvents can be potentially damaging to proteins. To overcome this problem new methods of microencapsulation are being suggested.

SOLVENT-FREE MICROENCAPSULATION TECHNOLOGY

The use of supercritical fluids as media for the formation of microparticles for therapeutic applications is a very recent development [4,28] The marked advantages that supercritical fluids offer over conventional microencapsulation processes include the mildness of the operating temperatures, the purity of the microcapsular products and the avoidance of organic solvents. Although the potential benefits of supercritical fluids are numerous for fragile drug microencapsulation, research in this area is still exploratory since a fundamental understanding of the topics has not been fully achieved.

A supercritical fluid is any substance the temperatures and pressure of which are simultaneously higher than the critical point values. The critical data for some of the common supercritical solvents are depicted in Table 1 for illustration purposes.

The most important property of a supercritical fluid is its large compressibility. This is because all fluids are infinitely compressible at the critical point. In the vicinity of the critical points, small changes in temperature or pressure generate substantial density variations. Supercritical fluids can be almost liquid-like in density and exhibit a viscosity that is intermediate between gas and liquid-like. Therefore, supercritical fluids

Table 1. Critical Temperature and Pressure of Selected Fluids

Substance	Critical Temperature (°C)	Critical Pressure (bar)
Oxygen	-119	50
Nitrogen	-147	34
Carbone dioxide	31	73
Ethane	32	48

are typically hundreds of times denser than gases at ambient conditions, but they are more compressible as previously mentioned.

There are many applications for supercritical fluids such as size reductions of pharmaceuticals, solvation or precipitation of pure compounds and polymer-drug microparticles (monolithic microcapsules) formation.[28] In the present case, it is intended to focus on the exploitation of the technique for microparticulate delivery systems formation only. For microcapsule formation it is the combination of liquid-like density and large compressiblity that is of interest. The large density means that supercritical fluids have solvent powers that are comparable with those of the same substances in the liquid state.[29] The high compressibility means that this solvent power is continuously adjustable between gas- and liquid-like extremes with small changes of pressure.

There are two routes to microcapsule formation with supercritical fluids: the rapid expansion of supercritical solutions (RESS), and the supercritical antisolvent (SAS) routes.[30]

Rapid Expansion of Supercritical Solution Technique (RESS)

Of the two routes to microcapsule formation with supercritical fluids, RESS[31] is used to form fine microparticles of substances that are appreciably soluble in a supercritical solvent (commonly CO_2). A schematic description of the RESS apparatus based on previous reports[32-34] is presented in Figure 5. The drug and the polymer are dissolved in a supercritical fluid at high pressure and precipitated by rapid decompression.

The RESS equipment consists of two main units, extraction (or dissolution) and precipitation. Prior to the dissolution of the drug and polymer, the pure solvent is pumped to the desired pressure and pretreated to extraction temperature by circulating through a preheater (consisting either of a coil or vessel) in a controlled temperature environment. In the precipitation unit, the supercritical solution is expanded across a capillary or an orifice as shown in Figure 5. However, many of the process variables affect the morphology of RESS powders in ways not yet fully understood. Thus, the morphology of the precipitate can be difficult to control and predict. In addition, the polymers used in controlled-delivery applications are not appreciably soluble in desirable supercritical fluids, such as CO_2. The solubility of low molecular weight poly(hydroxy acids) is typically less than 0.5% w/w. [32,35] Possible solutions to this problem are the use of cosolvents (Figure 5) to enhance the solubility, or the processing of low molecular weight polymers, which are suitable for shorter-release periods. Such cosolvents tend to be organic solvents (i.e., acetone) or fluorohalocarbons (i.e., $CHClF_2$), both less desirable than CO_2 from an environmental perspective.

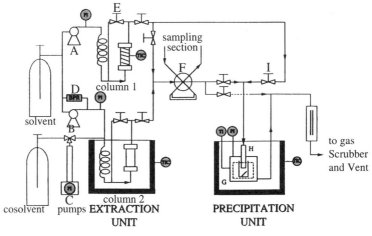

Figure 5. Schematic description of the rapid expansion of supercritical solution (RESS) apparatus (28).

Tom and Debenedetti[32] found that commercial L-PLA (Mw = 5000 - 6000: Mw/Mn = 0.2, where Mw and Mn are the respective weight-average and number-average polymer molecular weight) was soluble in supercritical CO_2 (0.05% w/w polymer) and in CO_2 with 1% w/w of acetone (0.37% w/w polymer).

These authors obtained 10- to 20- μm microspheres, and dendrites several hundred microns in size. The extreme sensitivity of particle morphology to small changes in process conditions is typical of experiments using orifices, from which virtually all of the expansion takes place in an uncontrolled free jet. Tom and Coll.[35] examined the effect of two types of devices: orifice (L/D <10) and capillaries (L/D >100) on microparticle morphology. The higher L/D ratio of capillaries relative to orifices leads to a larger density drop within the expansion device and, hence, to particle formation under more controlled conditions. Capillaries with inner diameters of 30 and 50 μm were used. When using the 30-μm-i.d. capillary with lengths ranging from 5 to 15 mm (167 < LD <500), a range of morphologies was produced by RESS (Figure 6). At high L/D ratios (500), only microparticles and agglomerates of microparticles were obtained, regardless of the preexpansion temperature (Figure 6a). At an L/D ratio of 350, microparticles were formed at high preexpansion temperatures, and microspheres were formed at low preexpansion temperatures (Figure 6b and 6c). The microspheres are most likely formed outside the capillary. Further evidence for the formation of microspheres outside the capillaries can be found in their sizes, which are often greater than the capillary's diameter (Figure 6d). The work of Tom and co-workers[35] with monodisperse L-PLA provided the first comprehensive experimental study on the effect of RESS process variables on microparticle and microsphere formation using biodegradable polymers.

Lovastatin, an anticholesterol drug was encapsulated within DL-PLA using the RESS technique.[36] Under appropriate experimental conditions, 36% lovastatin content microcapsules were obtained. Further research could lead to important advances in the production of injectable, controlled release monolithic microcapsule delivery systems. Work to date on these RESS processes for microencapsulation using biodegradable polymers appears to be promising.

However, to overcome the marked limitation of the low solvent power of supercritical fluids with moderate critical temperature (e.g. CO_2) toward high molecular polymers and many active molecules, the supercritical antisolvent processing (SAS) was proposed.

Supercritical Antisolvent Technique (SAS)

In SAS drug particles can be either dispersed or dissolved in the organic solvent in which the coating polymer has already been dissolved. The organic phase is then contacted with a supercritical fluid with a low affinity for the drug and polymer and appreciable mutual solubility with the organic phase. This results in organic solvent

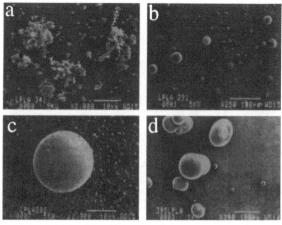

Figure 6. (a) L-PLA microparticles and (b-d) microspheres produced by RESS using capillaries as the expansion device (35).

expansion by the compressed antisolvent and organic solvent evaporation into the antisolvent phase (organic solvent extraction by the supercritical fluid).

The SAS apparatus shown in Figure 7 with appropriate modifications can be used to produce microparticulate delivery systems by contacting the organic phase and the supercritical fluid in a continuous, downflow cocurrent pattern.[36-38] The continuous SAS processing has been described by Knutson and Coll.[28] In this mode of operation, the organic solution is pumped from a feeding tank (12) to a nozzle (7) (20- to 300-μm i.d.) located at the top of the precipitator (5). The pressure drop of the organic solution across the nozzle is approximately 15 bar and is required for droplet formation. The nozzle discharges submillimeter liquid droplets into a cocurrently flowing supercritical stream, which also enters from the top of the precipitator. With the correct choice of solvent-antisolvent systems and operating conditions, the organic solvent is rapidly dissolved in the supercritical phase, leaving behind solute particles. In continuous operation, the organic solution and the supercritical fluid streams, the respective flow rates of which are 0.05-0.5 ml/min and 5-20 SLPM, are contacted for several hours in the 75-ml precipitator. By using CO_2 as the antisolvent, the typical operating pressure ranges from 75 to 135 bar. A wide range of solute concentrations in the organic phase has been used to date: 0.1 mg/ml (e.g., precipitation of insulin or catalase from 90% ethanol/10% DI water[35] to 50 mg/ml (e.g., precipitation of D,L-poly(lactic acid) from DMSO). The resulting particles are collected on glass slides (6) or on a filter (8) located before the precipitator outlet. System pressure is maintained through a series of backpressure regulators (4). The outlet stream is partially expanded in a depressurization vessel (10), where the organic solvent can be recovered. The antisolvent phase then passes through a flowmeter (11) and a dry test meter (13) after depressurization to atmospheric pressure. With only supercritical fluid flowing in the precipitator, this same flow pattern is used to evaporate the remaining organic solvent from the particles.

It should be pointed out that other terms were also used in the literature to describe the basic SAS process such as the precipitation with a compressed antisolvent (PCA)[39] or the aerosol solvent extraction system[40] depending either on the modifications or experimental conditions used.

Carbon dioxide has been the antisolvent of choice for use in protein processing owing to its nontoxicity, mild critical temperature, and environmental acceptability as a solvent. Investigations of protein powder formation by SAS have involved bovine insulin and bovine liver catalase.[36] Furthermore Tom and Coll.[36] precipitated both catalase and insulin, separately, from 90% ethanol/10% DI water solution, using compressed CO_2 antisolvent. In subsequent experiments Yeo and Coll.[37] used DMSO or DMF as the organic solvent, and produced biologically active insulin microparticles ranging in size from 1 to 4 μm for 90% of the population and less than 1 μm for the 10 remaining percent. In spite of the full insulin activity proven in animal experiments, recent results suggest significant disruption of the secondary structure of the SAS-

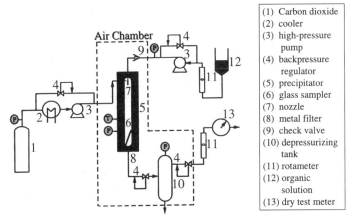

(1)	Carbon dioxide
(2)	cooler
(3)	high-pressure pump
(4)	backpressure regulator
(5)	precipitator
(6)	glass sampler
(7)	nozzle
(8)	metal filter
(9)	check valve
(10)	depressurizing tank
(11)	rotameter
(12)	organic solution
(13)	dry test meter

Figure 7. Schematic diagram of the experimental apparatus for batch and continuous SAS processing (37).

processed insulin.[41] Further research is necessary to ascertain whether this phenomenon occurs with proteins other than insulin.

Randolph et al.[39] precipitated L-PLA microparticles from methylene chloride, using near-critical and supercritical CO_2. In continuous-mode operation, the L-PLA-methylene chloride solution (0.3% w/w) polymer was sprayed into concurrently flowing CO_2 through an ultrasonic atomizer. This atomizer acted to break the polymer solution into droplets by means of a 120-kHz vibration. Polymer solution flow rates were adjusted between 0.02 and 2.0 ml/min, while the CO_2 flow rate was maintained at 151 SLPM. Continuous experiments were conducted at $36°$ and $40°C$. The suspended particle-methylene chloride-CO_2 system formed in the precipitator was expanded across a micrometering valve. The collection of dry samples on a glass slide placed at the micrometering valve outlet was possible because of the low boiling point of methylene chloride ($40°C$). Nearly spherical particles of PLA (0.61-1.4 μm size) were produced.

It was also shown that methylene chloride can be extracted from PLGA microcapsules down to undetectable levels by supercritical CO_2[4] or below the residual amounts (500 ppm) specified by the United States Pharmacopeia.[42] This opened up the possibility for using the SAS technique to successfully microencapsulate drug-biodegradable polymer systems particularly sensitive to heat, hydrolysis and pH. In fact this method is suited for the microencapsulation of proteins if they are not denatured by methylene chloride or DMSO under the SAS conditions.

The SAS technology has also been applied to the formation of tyrosine-derived polycarbonate and polyarylate microspheres. This series of pseudopoly(amino acids) are currently being studied as biodegradable material.[43]

Microspheres of a tyrosine-based polyarylate, poly(DTH sebacate), were obtained by the fast-batch expansion (20 bar/min) of a 5.5-mg/ml-DMSO solution with CO_2. At $35°C$, visible particle formation occurred at 73.4 bar. Microspheres, with an average particle diameter of approximately 200 μm, were sampled from a glass slide within the precipitator. However, SEM results suggest that these particles may be hollow.

A very different morphology results from the precipitation of a related pseudopoly(amino acid), poly(DTB carbonate) in a continuos, cocurrent SAS process. In a precipitator maintained at $35.1°C$ and 101 bar, a 6.1-mg/ml-polymer-DMSO solution was sprayed cocurrently into flowing CO_2. The polymer solution and CO_2 flow rates were 0.07 ml.min and 16 SLPM, respectively. The resulting microspheres, collected on a Teflon strip in the precipitator, are shown in Figure 8. The polymeric microspheres are typically smaller than 4 μm in diameter and show no indication of being hollow.

The inherent versatility of the SAS technique suggests a high likelihood for the successful microencapsulation of a variety of drug-bioerodible polymer systems. Work on SAS microencapsulation is in progress in various research laboratories.

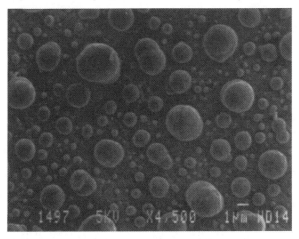

Figure 8. SEM photomicrographes of the pseudopoly (amino acid), poly (DTB carbonate), precipitate in continuous, countercurrent mode from DMSO using CO_2 as the supercritical antisolvent (T=35.1⁰ C; P=101 bar) (43).

To date, Müller and Fischer[44] have reported on the coprecipitation of drug and polymer by SAS techniques. The drug clonidine hydrochloride was microencapsulated in an L-PLA matrix from a solution of the drug and polymer in methylene chloride. The organic phase was contacted with CO_2 in continuous, countercurrent flow. Microspheres approximately 100 μm in diameter were formed at 100 bar and 60°C.

Recently, Bleich and Müller[45] reported the results of investigations regarding the incorporation of different model drugs into a biodegradable polymer carrier (poly-L-lactide) using the ASES process. The drugs such as hyoscine butylbromide, indomethacin, piroxicam and thymopentin were selected according to their polarity. Due to the miscibility of organic solvent (methylene chloride) and supercritical gas phase (CO_2), microparticles with residual organic solvents below 30 ppm are formed. The resulting microparticles were investigated with regard to particle formation, morphology, particle size, size distribution, and drug loading. With decreasing polarity of the incorporated drug, an increasing extraction occurs which lowers the drug loading of the microparticles. The extraction capacity of the gas phase depended on temperature and pressure which determined density and polarity of the gas. The obtained results showed that the production conditions have to be optimized for each drug/polymer combination. Totally non-polar drugs were completely extracted together with the organic solvent, however, polar drugs, and here especially peptides and proteins, are easy to incorporate with the ASES process.

More recently, Falk and Coll.[46] prepared poly (L-lactide) microspheres containing low molecular weight pharmaceutical agents using the PCA process with supercritical carbon dioxide as the antisolvent. Gentamycin, naloxone, and naltrexone were solubilized in methylene chloride using hydrophobic ion pairing (HIP) to stoichiometrically replace polar counter ions with an anionic detergent, aerosol OT (AOT, sodium bis-2-ethylhexyl sulfosuccinate). Through HIP complexation, solubilities in excess of 1 mg/ml were attainable in methylene chloride, allowing levels of direct incorporation that are not possible with other PCA approaches. The drug/polymer particles were spherical in shape and between 0.2 and 1.0 μm in diameter, as determined by scanning electron microscopy. At 37°C, the release of the ion-paired drugs into phosphate-buffered saline displayed minimal burst effects and exhibited release kinetics that were approximately linear with the square root of time, indicating matrix diffusion control of drug release. For gentamycin, linear release from the poly (L-lactide) microspheres was observed for more than 7 weeks, even at a drug loading of near 25% (w/w). Naltrexone exhibits similar release characteristics, although more drug was found on the surface of the microspheres. Conversely, rifampin, which was not ion-paired, was poorly encapsulated.

CONCLUSIONS

The application of supercritical fluids to the production of microparticulates of biomedical interest is a recent development, and the field is still in its infancy. The main advantage of using supercritical fluids are the purity of the products, which avoids or minimizes downstream processing; the mildness of the operating temperatures; and the simplicity of the processes. Both RESS and SAS techniques are compatible with the processing of sensitive recombinant proteins and peptides and may be useful for their encapsulation in biodegradable injectable controlled release microparticulate delivery systems.

Given the increasingly stringent regulations limiting the emission of organic solvents and aqueous waste streams, supercritical fluids, and in particular supercritical carbon dioxide, are attractive alternatives to conventional solvents. RESS and SAS are also more environmentally benign than actual organic solvent evaporation techniques. Even when organic solvents are mixed with supercritical CO_2, only small amounts are needed and can be easily recovered downstream by depressurization. The environmental and economic incentives for the development and improvement of carbon dioxide-based alternatives to existing technologies are strong. Research emphasis will probably focus in the near future on these promising microencapsulation techniques.

REFERENCES

1. R. Langer, 1990, New methods of drug delivery, *Science* **249**:1527-1533.
2. J.R. Nixon, 1976, *Microencapsulation*, Marcel Dekker, Inc., New York.
3. J.E. Vandergaer, 1974, *Microencapsulation, Processes and Applications*, Plenum Press, New York.
4. J.P. Benoit, H. Marchais, H. Roland and V. Van de Velde, 1996, Biodegradable microspheres: advances in production and technology, in:*Microencapsulation Methods and Industrial Applications*, (S. Benita, ed.), pp. 35-71, Marcel Dekker, Inc., New York.
5. H. Parmar, S.L. Lightman, L. Allen, R.H. Phillips, L. Edwards, and A.V. Schally, 1985, Randomised controlled study of orchidectomy vs long-acting DTrp6-LHRH microcapsules in advanced prostatic carcinoma, *Lancet II* **8466**:1201-1205.
6. J.M. Ruiz, B. Tissier, and J.P. Benoit, 1989, Microencapsulation of peptides: A study of the phase separation of poly (D,L-lactic acid-co-glycolic acid) copolymers 50/50 by silicone oil, *Int. J. Pharm.* **49**:69-77.
7. I. Heron, F. Thomas, M. Dero, J.R. Poutrain, S. Henane, F. Catus, and J.M. Kuhn, 1993, Traitement de l'acromegalie par une forme a liberation prolongee du lanreotide, un nouvel analogue de la somatostatine, *Presse Med.* **22**(1):526-531.
8. P. Orsolini, R.Y. Mauvernay, and R. Deghenghi, 1988, Procédé pour la microencapsulation par séparation de phases de substances médicamenteuses hydrosoluble, Swiss Patent 665 558 A5.
9. B.W. Muller, J. Bleich, and B. Wagenaar, 1993, Microparticle production without organic solvent, Proceedings of 9h International Symposium on Microencapsulation, Ankara, pp. 29-40.
10. J. Broadhead, S.K.E. Rouan, and C.T. Rhodes, 1992, The spray-drying of pharmaceuticals, *Drug Dev. Ind. Pharm.* **18**(11-12):1169-1206.
11. M. Montini, A. Pedroncelli, F. Tengattini, M. Pagani, D. Gianola, L. Cortesi, G. Pagani, and I. Lancranjan, 1993, Medical applications of intramuscularly administered bromocriptine microspheres, in: *Pharmaceutical Particulate Carriers, Therapeutic Applications*, (A. Rolland, ed.), Marcel Dekker, New York.
12. R. Schmiedel and J.K. Sandow, 1989, Microcapsule production containing soluble protein or peptide using mixture of poly:hydroxy-butyric acid and poly:lactide-co-glycolide, Eur. Patent 315875A.
13. M.S. Hora, R.K. Rana, J.H. Nunberg, T.R. Tice, R.M. Gilley and M.E. Hudson, 1990, Release of human serum albumin from poly(lactide-co-glycolide) microspheres, *Pharm. Res.* **7**:1190-1194.
14. R. Jalil and J.R. Nixon, 1990, Biodegradable poly(lactic) acid) and poly(lactide-co-glicolide) microcapsules: problems associated with preparative techniques and release properties, *J. Microencapsulation* **7**:297-325.
15. J.M. Ruiz, J.P. Busnel and J.P. Benoit, 1990, Influence of average molecular weights of poly(DL-lactic acid-co-glycolic acid) copolymers 50/50 on phase separation and in vitro drug release from microspheres, *Pharm. Res.* **7**:928-934.
16. J.P. McGee, S.S. Davis and D. O'Hagan, 1995, Zero order release of protein from poly(D,L-lactide-co-glycolide) microparticles prepared using a modified phase separation technique, *J. Control. Rel.* **34**:77-86.
17. W.W. Thompson, D.B. Anderson and H.L. Heiman, 1997, Biodegradable microspheres as a delivery system for rismorelin porcine, a porcine-growth-hormone-releasing-hormone, *J. Control. Rel.* **43**:9-22.
18. J.L. Cleland and A.J.S. Jones, 1996, Stable formulations of recombinant human growth hormone and inteferon-γ for microencapsulation in biodegradable microspheres, *Pharm. Res.* **13**:1464-1475.
19. Y. Ogawa, M. Yamamoto, H. Okada, T. Yashiki and T. Shimamoto, 1988, A new technique to efficiently entrap leuprolide acetate into microcapsules of polylactic acid or copoly(lactic/glycolic) acid, *Chem. Pharm. Bull.* **36**:1095-1103.
20. D.T. O'Hagan, H. Jeffery and S.S. Davis, 1994, The preparation and characterization of poly(lactide-co-glycolide) microparticles: III. Microparticle/polymer degradation rates and the in vitro release of a model protein, *Int. J. Pharm.* **103**:37-45.
21. C. Yan, J.H. Resau, J. Hewetson, M. Weset, W.L. Rill and M. Kende, 1994, Characterization and morphological analysis of protein-loaded poly(lactide-co-glycolide) microparticles prepared by water-in-oil-in-water emulsion technique, *J. Control. Rel.* **32**:231-241.

22. M.J. Alsonso, S. Cohen, T.W. Park, R.K. Gupta, G.R. Siber and R. Langer, 1993, Determinants of release of tetanus vaccine from polyester microspheres, *Pharm. Res.* **10**:945-953.

23. M.J. Blanco, F. Delie, E. Fattal, A. Tartar, F. Puisieux, A. Gulik and P. Couvreur, 1994, Characterization of V3 BRU peptide-loaded small PLGA microspheres prepared by a (w/o/w)/w emulsion solvent evaporation method, *Int. J. Pharm.* **111**:137-145.

24. H. Okada, Y. Doken, Y. Ogawa and H. Toguchi, 1994, Preparation of three-month depot injectable microspheres of leuprorelin acetate using biodegradable polymers, *Pharm. Res.* **11**:1145-1147.

25. I. Soriano, C. Evora and M. Llabrés, 1996, Preparation and evaluation of insulin-loaded poly(DL-lactide) microspheres using an experimental design, *Int. J. Pharm.* **142**:135-142.

26. H. Marchais, F. Boury, C. Damgé, J.-E. Proust and J.-P. Benoit, 1996, Formulation of bovine serum albumin loaded PLGA microspheres Influence of the process variables on the loading and in vitro release, *S.T.P. Pharma Sciences* **6**(6):417-423.

27. M. Morlock, H. Koll, G. Winter, T. Kissel, 1997, Microencapsulation of rh-erythropoietin, using biodegradable poly(D,L-lactide-co-glycolide): protein stability and the effects of stabilizing excipients, *Eur. J. Pharm. Biopharm.*, **43**:29-36.

28. B.L. Knutson, P.G. Debenedetti, 1996, Preparation of microparticulate using supercritical fluids, in: Microparticulate Systems for the Delivery of Proteins and Vaccines (S. Cohen and H. Bernstein, eds.), pp. 89-125, Marcel Dekker, Inc., New York.

29. S.K. Kumar and K.P. Johnston, 1988, Modelling the solubility of solids in supercritical fluids with density as independent variable, *J. Supercrit. Fluids* **1**:15.

30. P.G. Debenedetti, 1994, Supercritical fluids as particle formation media, *Supercritical Fluids--Fundamentals for Application* (E. Kiran and J.M.H. Levelt Sengers, eds.), *NATO ASI Ser. E* **273**:719.

31. J.W. Tom and P.G. Debenedetti, 1991, Particle formation with supercritical fluids--a review, *J. Aerosol. Sci.* **22**:555-584.

32. J. Tom and P.G. Debendetti, 1991, Formation of bioerodible polymeric microspheres and microparticles by rapid expansion of supercritical solutions, *Biotechnol. Prog.* **7**:403-411.

33. C.J. Chang and A.D. Randolph, 1989, Precipitation of microsize organic particles from supercritical fluids, *AIChE J.* **35**:1876-1882.

34. D.W. Matson, J.L. Fulton, R.C. Petersen, and R.D. Smith, 1987, Rapid expansion of supercritical fluid solutions: Solute formation of powders, thin films and fibers, *Ind. Eng. Chem. Res.* **26**:2298.

35. J.W. Tom, P.G. Debenedetti, and R. Jerome, 1994, Precipitation of poly(l-lactic acid) and composite poly(l-lactic acid)-pyrene particles by rapid expansion of supercritical solutions, *J. Supercrit. Fluids* **4**:429.

36. J.W. Tom, G. Lim, P.G. Debenedetti, and R.K. Prud'homme, 1992, Applications of supercritical fluids in controlled release of drugs, *Supercritical Fluid Engineering Science ACS Symp. Ser. 514* (E. Kiran and J.F. Brennecke, eds.), American Chemical Society, Washington, DC, pp.238-257.

37. S.-D. Yeo, G.-B. Lim, P.G. Debenedetti, and H. Bernstein, 1993, Formation of microparticulate protein powders using a supercritical fluid antisolvent, *Biotechnol. Bioeng.* **41**:341-346.

38. S.-D. Yeo, P.G. Debenedetti, M. Radosz, and H.-W. Schmidt, 1993, Supercritical antisolvent (SAS) process for substituted aromatic polyamides: Phase equilibrium and morphology study, *Macromolecules* **26**:6207.

39. T.W. Randolph, A.D. Randolph, M. Mebes, and S. Yeung, 1993, Sub-micrometric sized biodegradable particles of poly(L-lactic acid) via the gas antisolvent spray precipitation process, *Biotechnol. Prog.* **9**:429-435.

40. J. Bleich, B.W. Müller, and W. Waβmus, 1993, Aerosol solvent extraction system - a new microparticle production technique, *Int. J. of Pharm.* **97**:111-117.

41. S.-D. Yeo, P.G. Debenedetti, S.Y. Patro, and T.M. Przybycien, 1994, Second structure characterization of microparticulate insulin powders, *J. Pharm. S.* **83**:1651-1656.

42. F. Ruchatz, P. Kleinebudde, and B.W. Müller, 1997, Residual solvents in biodegradable microparticles. Influence of process parameters on the residual solvent in microparticles produced by the aerosol solvent extraction system (ASES) process, *J. Pharm. Sci.* **86**:101-105.

43. S.I. Ertel and J. Kohn, 1994, Evaluation of a series of tyrosine-derived polycarbonates as degradable biomaterials, *J. Biomed. Mater. Res.* **28**:919-930.

44. B.W. Muller and W. Fisher, Manufacture of sterile sustained-release drug formulations using liquified gas, W. Germany Patent 3,744,329, July 6, 1989.

45. J. Bleich and B.W. Müller, 1996, Production of drug loaded microparticles by the use of supercritical gases with the Aerosol Solvent Extraction System (ASES) process, *J. Microencapsulation* **13**:131-139.

46. R. Falk, T.W. Randolph, J.D. Meyer, R.M. Kelly, and M.C. Manning, 1997, Controlled release of ionic compounds from poly (L-lactide) microspheres produced by precipitation with a compressed antisolvent, *J. Control. Rel.* **44**:77-85.

RECENT ADVANCES IN NANOPARTICLES AND NANOSPHERES

Jörg Kreuter

Institut für Pharmazeutische Technologie
Johann Wolfgang Goethe-Universität
Biozentrum
Marie-Curie-Str. 9
60439 Frankfurt am Main

INTRODUCTION

Nanoparticles were first developed around 1970. They were initially devised as carriers for vaccines and anticancer drugs. In order to enhance tumor uptake the strategy of drug targeting was employed, and as a first important step research focussed on the development of methods to reduce the uptake of the nanoparticles by the cells of the reticuloendothelial system (RES). Simultaneously, the use of nanoparticles for ophthalmic and oral delivery was investigated.

MACROPHAGE TARGETING FOR AIDS THERAPY

Recently, research starts to take advantage of just these properties of nanoparticles to be easily taken up by cells of the RES. In the case of AIDS, for instance, the macrophages of the RES represent one of the most important therapeutic targets:

In addition to the CD4[+] T lymphocytes, these cells play a decisive role as a reservoir for the human immunodeficiency virus (HIV). In tissues such as the lung and the brain, HIV is located primarily in macrophage-like cells (i.e. alveolar macrophages and microglia, respectively)[1, 2]. In contrast to CD4[+] T cells, in which HIV replication is proliferation dependent and finally leads to cell death, macrophages in a mature nonproliferating but immunologically active state can be productively infected with HIV type 1 (HIV-1) and HIV-2[3 - 5]. Altered cellular functions in the macrophage population may contribute to the development and clinical progression of AIDS. A large proportion of AIDS patients shows severe HIV encephalitis with diffuse neuronal damage which is thought to be mediated by viral proteins and/or factors with neurotoxic activity (i.e. cytokines). This can occur through proinflammatory cytokines or by HIV-1-specific proteins secreted from cells of the mononuclear phagocyte system, including brain macrophages, microglia, and multinucleated giant cells, which have been shown to be productively infected with HIV[6-8].

There is evidence that, apart from having a function in the pathogenesis of the disease, cells of the macrophage lineage are vectors for the transmission of HIV. The placental macrophage is likely to be the primary cell type responsible for vertical (maternofetal) transmission of HIV[9]. For mucosal transmission it was found that an important property of the transmitted HIV variant is its ability to infect macrophages[10]. The phenotypic characterization of virus populations which were transmitted sexually or vertically revealed a selection of variants with a predominant tropism for macrophages in the recipient[11]. Because of the important role of cells of the monocyte/macrophage (Mo/Mac) lineage in the pathogenesis of HIV, fully effective anti-HIV therapy must reach Mo/Mac in addition to other target cells[8].

Consequently, nanoparticles without surfactant coating leading to stealth properties may represent promising drug carriers for this disease. Therefore, the potential on nanoparticles to deliver antiviral drugs to HIV-infected and uninfected macrophages was investigated by us[12, 13]. Indeed, it was found in in-vitro cell cultures of human monocytes/macrophages (Mo/Mac) that the uptake of nanoparticles by HIV-1-infected cells was about 30% higher after 7 days of culture and about 65% after 21 days than by uninfected cells. Similar results were observed with macrophages obtained from AIDS patients, depending on the state of disease. The uptake by these cells was also dependent on the surface coating with different surfactants (Table 1). Interestingly, despite their possible stealth properties these surfactants did not reduce the in-vitro uptake of the nanoparticles: poloxamer 338 had no influence and poloxamer 188 even increased the uptake[12, 13].

Table 1. Influence of surfactant on phagocytosis of PBCA nanoparticles by human MAC [reproduced from Schäfer et al., 13]

	Phagocytosed NP (μg/ml ± SD)[a]		
	Uncoated NP	Pluronic F108	Pluronic F68
IFN[b]	8.6±2.0	9.5±1.8	13.0±2.9
Control	6.4±0.3	7.1±1.4	9.8±2.6

[a] Mean values of three parallel cultures are presented.
[b] The cells were stimulated with IFN-gamma for 24 hr (200 U/ml).

The following investigation of the in-vitro inhibition of the development of infection of the above human macrophage cultures by HIV-1 using the antiviral drugs azidothymidin (AZT) and dideoxycytidine (ddC), however, revealed no difference between free and nanoparticle-bound drugs[14]: nanoparticle-bound AZT or ddC retained their activity but were not superior to free drug. The situation was totally different with the antiviral protease inhibitor saquinavir[8]. Here a 10-fold increase in efficacy of nanoparticle-bound drug over free drug was observed. The difference in these results probably is due to the fact that AZT and ddC easily penetrate into cells whereas saquinavir does not, and therefore requires the mediation of cell uptake by the nanoparticles.

However, it has to be considered that in tissue cultures a homologous cell population exists, which is totally different to the situation in the human or animal body: the body consists of a huge number of different cell types. Macrophages represent only a small percentage of the total number of cells. As a consequence, it was conceivable that even with AZT or ddC nanoparticles still may achieve a better targeting to theses cells despite their inability to show a better antiviral efficacy of these drugs in-vitro in the homologous cell cultures. Indeed that was observable: Löbenberg and Kreuter[15] bound ^{14}C-labeled AZT to nanoparticles. After i.v. injection of nanoparticle-bound drug the concentrations of the AZT-label in the organs that were rich in macrophages, such as the liver (Fig. 1), were up to 18 times higher than with an AZT solution. Likewise, after oral administration the nanoparticle formulation delivered the AZT-label more efficiently to the RES than the aqueous solution. In addition, the blood concentration was significantly higher after oral administration of nanoparticles. Autoradiographic investigations using radioluminography supported the above results[16]. The radioactivity in the liver, lung, and spleen, organs with a high number of macrophages, was much higher than in other parts of the body with the nanoparticle formulation. The above organs showed a spotted appearance of the radioactivity that is typical for the accumulation in macrophages after i.v. as well as after oral administration of the drug bound to the particles. In contrast, after administration of the solution the radioactivity in the above organs was homogeneously distributed and much lower. These results show that the nanoparticles seemed to have reached their target and that nanoparticles may represent very promising delivery system for AIDS therapy.

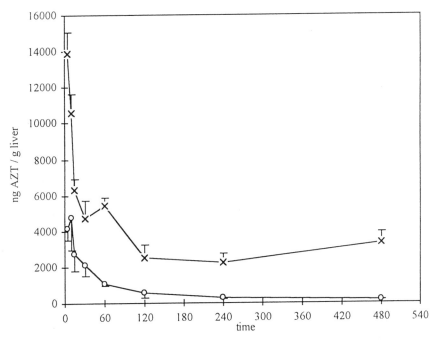

Figure 1. ^{14}C-AZT-label concentrations in the liver of rats after i.v. injection of an AZT solution o, or of AZT bound to poly/butyl cyanoacrylate) nanoparticles, x. (From Löbenberg and Kreuter (1996), reproduced with kind permission of the copyright holder, Mary Ann Liebert, Inc., Lachmont, N.Y., USA.)

Another new interesting area of application of nanoparticles is brain targeting. Tröster et al.[17 - 19] already observed that besides surfactants that achieved a prolonged circulation times of the particles, such as poloxamer 338 and poloxamine 908 and 1508, other surfactants, for instance some of the polysorbates, may be as efficient or even more efficient in increasing the concentrations in organs that do not belong to the RES. Tröster et al.[17] at that time assumed that the coating of the nanoparticles with the surfactants may have led to an accumulation of the nanoparticles in the brain blood vessels possibly by an increased adhesion to the blood vessel wall, but these authors did not believe that in the brain the particles were taken up by or transported across the endothelial cells lining these vessels. Later, however, Borchard et al.[20] observed an increased uptake of surfactant-coated poly(methyl methacrylate) nanoparticles by bovine brain microvessel endothelial cell monolayers in-vitro. The highest and fasted uptake (>300% compared to uncoated controls) was observed after coating with polysorbate 80. Poloxamer 407 also yielded a very high uptake that was, however, delayed by several hours. Poloxamers 184 and 188, polysorbate 80 and polyoxyethylen-23-laurylether showed an intermediate uptake enhancement (>100%) whereas poloxamer 338 and 908 only yielded a minimal insignificant uptake enhancement.

These findings were later supported by in-vivo experiments. These in-vivo experiments with fluorescent polysorbate 80-coated fluorescence isothiocyanate dextrane 70000-labeled poly(butyl cyanoacrylate) nanoparticles in mice[21] indicated using fluorescence and electron microscopy that the particles were taken up by the endothelial cells lining the brain blood vessels and showed transport of fluorescence into the Punkinje Cells of the brain 45 min after injection. No uptake occurred without polysorbate 80-coating. In addition, a strong analgesic effect again exhibiting a maximum after 45 min was observed in mice by the tail-flick test following i.v. injection of the polysorbate 80-coated poly(butyl cyanoacrylate) nanoparticles that were before loaded with dalargin (Fig. 2). Dalargin is a hexapeptide leu-enkephalin analogue with good blood stability that normally does not cross the blood-brain-barrier (BBB) after i.v. injection but possesses analgesic activity after direct intraventricular brain injection. Indeed, in the above experiments by Kreuter et al.[21] and by Alyautdin et al.[22] an intravenously injected dalargin solution as well as other controls showed no analgesia. The controls included empty nanoparticles, polysorbate 80 alone, dalargin plus polysorbate 80, dalargin bound to the nanoparticles but without polysorbate 80, and the simple mixture of all three components, dalargin, nanoparticles and polysorbate 80. In this latter mixture these three components were mixed immediately before injection, whereas in the analgetically active preparations dalargin was bound to the nanoparticles by incubation under stirring for 3 hours followed by the adsorption of polysorbate 80 by another 30 min. The analgesic effect achieved with this preparation was dose-dependent and it was totally blocked by pretreatment with naloxone, an opiod receptor antagonist, given i.v. 10 min before injection of the polysorbate 80 nanoparticle preparation. These experiments with the tail-flick test were later confirmed by Schröder and Sabel[23] using the hot plate test.

The observed analgesia with dalargin bound to nanoparticles depended on the surfactant employed: Only polysorbate 20, 40, 60, and 80 achieved an analgesic effect after overcoating of the dalargin nanoparticles and i.v. injection[24]. Polysorbate 80 enabled the highest induction of analgesia at both doses of dalargin employed, i.e. 7 mg/kg and 10 mg/kg. All other surfactants including poloxamer 184, 188, 388, and 407, poloxamine 908, polyoxyethylen-23-laurylether, Cremophor®EL and RH 40, and polysorbate 81 and 85 exhibited no effect. This shows that the transport of dalargin cannot be due to a simple surfactant effect on the endothelial cells or on the tight junctions since most of the surfactants were inactive. It might be argued that the polysorbates may interact much

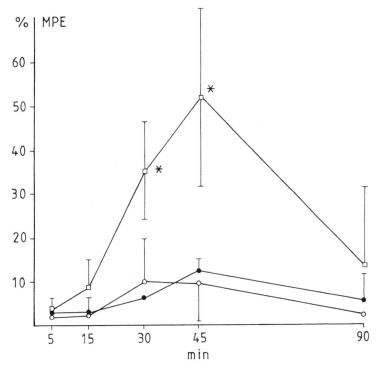

Figure 2. Maximal possible analgesic effect (MPE) determined by the tail-flick test after i.v. injection of one of the following preparations to mice: O, 10 mg/kg dalargin phosphate buffered saline, pH 7.2 (PBS); ●, a mixture of 10 mg/kg dalargin, poly(butyl cyanoacrylate) nanoparticles and polysorbate 80, mixed immediately before injection; □, 7.5 mg/kg dalargin adsorbed to poly(butyl cyanoacrylate) nanoparticles overcoated with polysorbate 80 + statistically significant difference (p < 0.05). (From Kreuter et al.[21], reproduced with kind permission of the copyright holder, Elsevier Science, Amsterdam, Netherlands)

stronger with the nanoparticles than the other surfactants. However, experiments in our laboratory have shown that this is not the case: On the contrary, poloxamine 908 interacts much stronger with the nanoparticles and influences the body distribution at much lower concentrations than polysorbate 80 after intravenous injection. It has to be noted that none of the „stealth" surfactants, poloxamer 338 nor poloxamine 908 exhibited any effect although both had achieved a high brain concentration of the nanoparticles in the study of Tröster et al.[17]. Also poloxamer 407 that was very active in the cell culture work of Borchard et al.[20] showed no effect in the above-mentioned study of Kreuter et al.[24]. It is possible, that the high brain concentrations of the nanoparticles observed after coating with poloxamer 338 and poloxamine 908 by Tröster et al.[17] are a reflection of their high blood concentrations but that these surfactants do not enable an interaction with the endothelial cells as already indicated by the in-vitro results by Borchard et al.[20]. Poloxamer 407-coated particles may be able to interact with endothelial cells, but because of the very slow onset of this process with this surfactant poloxamer 407 may not have led to a relevant response in-vivo or only after very extended times, i.e. after the observation interval.

The binding of other drugs to nanoparticles and overcoating with polysorbate 80 also led to pharmacologic responses. Loperamid too is an opiod that does not lead to central effects because it cannot cross the blood-brain-barrier and, therefore, is used perorally against diarrhea. This drug also became centrally active after binding to nanoparticles overcoated with polysorbate 80[25]. In the same manner tubocurarine, a muscle relaxant, induced spikes in the EEC in brain perfusion studies[26]. Again the controls, uncoated tubocurarine-loaded nanoparticles or a mixture of the drug with polysorbate 80, were inactive in enabling a passage of this drug across the blood-brain-barrier, and, consequently, the EEC remained normal.

Another example is doxorubicin. This drug was bound to poly(butyl cyanoacrylate) nanoparticles, and the blood plasma, liver, spleen, lung, heart, and kidney and brain concentrations were measured after i.v. injection to rats at a dose of 5 mg doxorubicin/kg (unpublished data). No significant differences were found between a solution of doxorubicin in saline and a doxorubicin solution in saline containing 1% polysorbate 80. Coating of the doxorubicin-loaded nanoparticles with this surfactant enhanced the plasma concentrations significantly and led to prolonged circulation times (Table 2). Both nanoparticle preparations, i.e. doxorubicin bound to nanoparticles without or with coating with polysorbate 80, reduced the heart concentration very significantly (Table 3). This observation is in excellent agreement with earlier data by Couvreur et al[27, 28].

However, in the brain only the polysorbate 80-coated nanoparticles achieved detectable concentrations (Table 4). The brain concentrations appeared rather late, after about 2 to 4 hours, but they were very considerable (>6µg/g). This level is more than 60 times above the detection limit whereas all other preparations were below the detection limit all the times. This result shows that this delivery system, polysorbate 80-coated nanoparticles, holds great promise for the delivery of antitumor drugs to the brain and may open up totally new possibilities in the treatment of brain cancers.

As already mentioned above the mechanism of the delivery seems to be endocytotic uptake of the drug-loaded nanoparticles by the endothelial cells lining the blood vessels of the brain. Endocytosis seems to be mediated by certain substances that are adsorbed from the blood after injection of the surfactant-coated particles[28]. Accordingly, the adsorption pattern of serum components after incubation of nanoparticles in serum was highly surfactant specific[29, 30]. No increased uptake of inulin was observed if the inulin was

Table 2. Doxorubicin concentrations [µg/g organ] in plasma after i.v. injection to rats [5 mg/kg] determined by HLPC

	10 min	1 h	2 h	4 h	6 h	8 h
DX	2.40±0.09	0.41±0.08	0.17±0.08	0.07±0.01	0.04±0.01	0.03±0.01
DX-80	2.70±0.12	0.37±0.04	0.15±0.02	0.05±0.01	0.05±0.01	0.03±0.00
DX-NP	1.68±0.07	0.21±0.03	0.10±0.04	0.08±0.01	0.09±0.01	0.07±0.00
DX-NP-80	3.15±0.42	0.89±0.10	0.21±0.07	0.09±0.01	0.09±0.01	0.06±0.08

DX   = Doxorubicin [5 mg/kg] in saline
DX-80   = Doxorubicin [5 mg/kg] in 1% polysorbate 80/saline solution
DX-NP   = Doxorubicin [5 mg/kg] bound to nanoparticles
DX-NP-80   = Doxorubicin [5 mg/kg] bound to nanoparticles and overcoated with 1% polysorbate 80

Table 3. Doxorubicin concentrations [µg/g organ] in heart after i.v. injection to rats [5 mg/kg] determined by HPLC

	10 min	1 h	2 h	4 h	6 h	8 h
DX	0.74±0.06	4.27±0.20	6.57±0.55	4.00±0.37	3.30±0.24	1.10±0.05
DX-80	1.10±0.35	8.40±0.62	5.73±0.50	2.10±0.30	0.75±0.08	0.40±0.05
DX-NP	0.28±0.04	0.30±0.05	<0.1	<0.1	<0.1	<0.1
DX-NP-80	0.42±0.07	<0.1	<0.1	<0.1	<0.1	<0.1

DX   = Doxorubicin [5 mg/kg] in saline
DX-80   = Doxorubicin [5 mg/kg] in 1% polysorbate 80/saline solution
DX-NP   = Doxorubicin [5 mg/kg] bound to nanoparticles
DX-NP-80   = Doxorubicin [5 mg/kg] bound to nanoparticles and overcoated with 1% polysorbate 80

Table 4. Doxorubicin concentrations [µg/g organ] in brain after i.v. injection to rats [5 mg/kg] determined by HPLC

	10 min	1 h	2 h	4 h	6 h	8 h
DX	<0.1	<0.1	<0.1	<0.1	<0.1	<0.1
DX-80	<0.1	<0.1	<0.1	<0.1	<0.1	<0.1
DX-NP	<0.1	<0.1	<0.1	<0.1	<0.1	<0.1
DX-NP-80	<0.1	2.20±0.49	6.13±0.27	6.20±0.80	2.67±0.18	1.10±0.07

DX	= Doxorubicin [5 mg/kg] in saline
DX-80	= Doxorubicin [5 mg/kg] in 1% polysorbate 80/saline solution
DX-NP	= Doxorubicin [5 mg/kg] bound to nanoparticles
DX-NP-80	= Doxorubicin [5 mg/kg] bound to nanoparticles and overcoated with 1% polysorbate 80

injected shortly after injection of the nanoparticles (unpublished). This indicates that the so-called inulin spaces were not significantly increased and that the tight junctions of the brain blood vessel endothelium were not opened.

Nevertheless, besides endocytosis another possibility exists for the increased brain uptake of the above drugs, namely inactivation of the P-glycoprotein efflux pump. This glycoprotein is present in the brain endothelial cells[31] and is, for instance, also responsible for the multidrug resistance which represents a major obstacle to cancer chemotherapy[32-34]. Surfactants including polysorbate 80 were shown to inhibit this efflux system and to reverse multidrug resistance. However, as mentioned above, addition of polysorbate 80 to the drug solutions was totally inefficient. On the other hand, it is conceivable that this surfactant or the other active polysorbates, of course, may be delivered more efficiently to the brain endothelial cell if they were adsorbed to the nanoparticle surface. This could explain why polysorbate-coated nanoparticles provided high brain concentrations in contrast to simple surfactant solutions. Nevertheless, we believe that the latter mechanism, i.e. blockage of the efflux system by polysorbate, may contribute to the higher brain/drug uptake mechanism but that induction of the endocytotic uptake may play an equal or much larger role. The reason for this assumption is the observation that in contrast to other organs significant brain concentrations were obtained only after some time. Such a delayed response is typical for a delivery mediated by the polysorbate-coated nanoparticles[21, 26] and seems to be a reflection of the time-consuming process of endocytosis.

A third novel area of interest is the binding of oligonucleotides to nanoparticles[35, 36]. These types of drugs due to their size and nature have major difficulties to reach their target cells. Nanoparticles may be useful for the targeting of oligonucleotiedes to these cells.

REFERENCES

1. J.A. Levy, 1993, Pathogenesis of human immunodeficiency virus infection. *Microbiol. Rev.* **57**: 183 - 289.
2. R.A. Weiss, 1993, How does HIV cause AIDS? *Science* **260**: 1273 - 1279.
3. S. Gartner, P. Markovitis, D.M. Markovitz, M.H. Kaplan, R.C. Gallo, and M. Popovic, 1986, The role of mononuclear phagocytes in HTLV-III/LAV infection. *Science* **223**: 215 - 219.
4. H. Kühnel, H. von Briesen, U. Dietrich, M. Adamski, D. Mix, L. Biesert, R. Kreutz, A. Immelmann, K. Henco, C. Meichsner, R. Andreesen, H. Gelderblom, and H. Rübsamen-Waigmann, 1989, Molecular cloning of two West African human immunodeficiency virus type 2 isolates that replicate well in macrophages: a Gambian isolate, from a patient with neurologic acquired immunodeficiency syndrome, and a highly divergent Ghanian isolate. *Proc. Natl. Acad. Sci. USA* **86**: 2383- 2387.
5. H. von Briesen, R. Andreesen, R. Esser, W. Brugger, C. Meichsner, K. Becker, and H. Rübsamen-Waigmann, 1990, Infection of monocytes/macrophages by HIV in vitro. *Res. Virol.* **141**: 225 - 231.

6. R. Esser, H. von Briesen, W. Brugger, M. Ceska, W. Glienke, S. Müller, A. Rehm, H. Rübsamen-Waigmann, and R. Andreesen, 1991, Secretory repertoire of HIV-infected human monocytes/macrophages. *Pathobiology* **59**: 219 - 222.

7. H.E. Gendelman, S.A. Lipton, M. Tardieu, M.I. Bukrinsky, and H.S.L.M. Nottet, 1994, The neuropathogenesis of HIV-1 infection. *J. Leukocyte Biol.* **56**: 389 - 398.

8. A.R. Bender, H. von Briesen, J. Kreuter, I.B. Duncan, and H. Rübsamen-Waigmann, 1996, Efficiency of nanoparticles as a carrier system for antiviral agents in human immunodeficiency virus-infected human monocytes/macrophages in vitro. *Antimicrob. Agents Chemother.* **40**: 1467-1471.

9. K.A. McGann, R. Collmann, D.L. Kolson, F. Gonzalez-Scarano, G. Coukos, C. Coutifaris, J.F. Strauss, and N. Nathanson, 1994, Human immunodeficiency virus type 1 causes productive infection of macrophages in primary placental cell cultures. *J. Infect. Dis.* **169**: 746 - 753.

10. G. Milman, and O. Sharma, 1994, Mechanisms of HIV/SIV mucosal transmission. *AIDS Res. Hum. Retroviruses* **10**: 1305 - 1312.

11. A.B. Van't-Wout, N.A. Kootstra, G.A. Mulder-Kampinga, N. Albrecht-van-Lent, H.J. Scherpbier, J. Veenstra, K. Boer, R.A. Coutinho, F. Miedema, and H. Schuitemaker, 1994, Macrophage-tropic variants initiate human immunodeficiency virus type 1 infection after sexual, parenteral, and vertical transmission. *J. Clin. Invest.* **94**: 2060 - 2067.

12. V.M. Schäfer, H. von Briesen, S.D. Tröster, R. Andreesen, J. Kreuter, and H. Rübsamen-Waigmann, 1991, How to get antiviral drugs into HIV-infected cells? Phagocytosis of nanoparticles by human macrophages. *Microbiologist* **2**: 117 - 121.

13. V. Schäfer, H. von Briesen, R. Andreesen, A.-M. Steffan, C. Royer, S. Tröster, J. Kreuter, and H. Rübsamen-Waigmann, 1992, Phagocytosis of nanoparticles by human immunodeficiency virus (HIV)-infected macrophages: A possibility for antivral drug targeting. *Pharm. Res.* **9**: 541 - 546.

14. A. Bender, V. Schäfer, A.M. Steffan, C. Royer, J. Kreuter, H. Rübsamen-Waigmann, and H. v. Briesen, 1994, Inhibition of HIV in vitro by antiviral drug-targeting using nanoparticles. *Res. Virol.* **145**: 215 - 220.

15. R. Löbenberg, and J. Kreuter, 1996, Macrophage targeting of azidothymidine: A promising strategy for AIDS therapy. *AIDS Res. Human Retrovir.* **12**: 1709 - 1715.

16. R. Löbenberg, J. Maas, and J. Kreuter, (in press), Improved body distribution of ^{14}C-labelled AZT bound to nanoparticles in rats determined by radioluminography. *J. Drug Target.*

17. S.D. Tröster, U. Müller, and J. Kreuter, 1990, Modification of the body distribution of poly(methyl methacrylate) nanoparticles in rats by coating with surfactants. *Int. J. Pharm.* **61**: 85 - 100.

18. S.D. Tröster, and J. Kreuter, 1992, Influence of the surface properties of low contact angle surfactants on the body distribution of ^{14}C-poly(methyl methacrylate) nanoparticles. *J. Microencapsul.* **9**: 19 - 28.

19. S.D. Tröster, K. H. Wallis, R.H. Müller, and J. Kreuter, 1992, Correlation of the surface hydrophobicity of ^{14}C-poly(methyl methacrylate) nanoparticles to their body distribution. *J. Controlled Rel.* **20**: 247 - 260.

20. G. Borchard, K.L. Audus, F. Shi, and J. Kreuter, 1994, Uptake of surfactant-coated poly(methyl methacrylate)-nanoparticles by bovine brain microvessel endothelial cell monolayers. *Int. J. Pharm.* **110**: 29 - 35.

21. J. Kreuter, R.N. Alyautdin, D.A. Kharkevich, and A.A. Ivanov, 1995, Passage of peptides through the blood-brain barrier with colloidal polymer particles (nanoparticles). *Brain Res.* **674**: 171 - 174.

22. R. Alyautdin, D. Gothier, V. Petrov, D. Kharkevich, and J. Kreuter, 1995, Analgesic activity of the hexapeptide dalargin adsorbed on the surface of polysorbate 80-coated poly(butyl cyanoacrylate) nanoparticles. *Europ. J. Pharm. Biopharm.* **41**: 44 - 48.

23. U. Schröder, and B.A. Sabel, 1996, Nanoparticles, a drug carrier system to pass the blood-brain-barrier, permit central analgesic effects of i.v. dalargin injections. *Brain Res.* **710**: 121 - 124.

24. J. Kreuter, V.E. Petrov, D.A. Kharkevich, and R.N. Alyautdin, (in press), Influence of the type of surfactant on the transport of the peptide dalargin across the blood-brain barrier using surfactant-coated nanoparticles. *J. Controlled Rel.*

25. R.N. Alyautdin, V. E. Petrov, K. Langer, A. Berthold, D.A. Kharkevich, and J. Kreuter, 1997, Delivery of loperamide across the blood-brain barrier with polysorbate 80-coated polybutylcyanoacrylate nanoparticles. *Pharm. Res.* **14**: 325 - 328.

26. R.N. Alyautdin, E.B. Tezikov, P. Ramge, D.A. Kharkevich, D.J. Begley, and J. Kreuter, (in press), Significant entry of tubocurarine into the brain of the rat is enabled by adsorption to polysorbate 80-coated polybutyl-cyanoacrylate nanoparticles: An in situ brain perfusion study. *J. Microencapsul.*

27. P. Couvreur, B. Kante, L. Grislain, M. Roland, and P. Speiser, 1982, Toxicity of polyalkylcyanoacrylate nanoparticles II: Doxorubicin-loaded nanoparticles. *J. Pharm. Sci.* **71**: 790 - 792.

28. P. Couvreur, L. Grislain, V. Lenaerts, F. Brasseur, P. Guiot, and A. Biernacki, 1986, Biodegradable polymeric nanoparticles as drug carrier for antitumor agents; in: P. Guiot and P. Couvreur (eds.), *Polymeric Nanoparticles and Microspheres*, pp. 27 - 93, Boca Raton: CRC Press.

29. J. Kreuter, 1983, Evaluation of nanoparticles as drug-delivery systems III: Materials, stability, toxicity, possibilities of targeting, and use. *Pharm. Acta. Helv.* **58**: 242 - 250.

30. D.F. Blunk, Hochstrasser, B.W. Müller, and R.H. Müller, 1993, Differential adsorption: Effect of plasma protein adsorption patterns on organ distribution of colloidal drug carriers. *Proceed. Intern. Symp. Control. Rel. Bioact. Mater.* **20**: 256 - 257.

31. D.F. Blunk, Hochstrasser, J.-C. Sanchez, B.W. Müller, and R.H. Müller, 1993, Colloidal carriers for intravenous drug targeting: Plasma protein adsorption patterns on surface-modified latex particles evaluated by two-dimensional polyacrylamide gel electrophoresis. *Electrophoresis* **14**: 1382 - 1387.

32. C. Cordon-Cardo, J.P. O'Brien, D. Casals, L. Rittmann-Grauer, J.L. Biedler, M.R. Melamed, and J.R. Bertino, 1989, Multidrug resistance gene (P-glycoprotein) is expressed by endothelial cells at blood brain barrier sites. *Proc. Natl. Acad. Sci., USA* **86**: 695 - 698.

33. D.M. Woodcock, S. Jefferson, M.E. Linsenmeyer, P.J. Crowther, G.M. Chojnowski, B. Williams, and I. Bertoncello, 1990, Reversal of multidrug resistance phenotype with Cremophor EL, a common vehicle for water-insoluble vitamins and drugs. *Cancer Res.* **50**: 4199 - 4203.

34. D.M. Woodcock, M.E. Linsenmeyer, G. Chojnowski, A.B. Kriegler, V. Nink, L.K. Webster, and W.H. Sawyer, 1992, Reversal of multidrug resistance by surfactants. *Br. J. Cancer* **66**: 62 - 68.

35. T. Zordan-Nudo, V. Ling, Z. Liv, and E. Georges, 1993, Effects of nonionic detergents on P-glycoprotein drug binding and reversal multidrug resistance. *Cancer Res.* **53**: 5994 - 6000.

36. C. Chavany, T. Saison-Behmoaras, T. Le Doan, F. Puisieux, P. Couvreur, and C. Hélène, 1994, Adsorption of oligonucleotides onto polyisohexylcyanoacrylate protects them against nucleases and increases their cellular uptake. *Pharm. Res.* **11**: 1370 - 1378.

37. H.-P. Zobel, J. Kreuter, D. Werner, C.R. Noe, G. Kümel, and A. Zimmer, (in press), Cationic polyhexylcyanoacrylate nanoparticles as carriers for antisense-oligonucleotides. *Antisense Nucl. Acid. Drug Per.*

EVALUATION AND FORMULATION OF BIODEGRADABLE LEVODOPA MICROSPHERES USING 3² FACTORIAL DESIGN

Betül Arıca, H. Süheyla Kaş, and A. Atilla Hıncal

Hacettepe University
Faculty of Pharmacy
Department of Pharmaceutial Technology
06100 Sıhhiye-Ankara
TURKEY

INTRODUCTION

The main neurochemial characteristic of the Parkinson's disease is a marked degeneration of the nigrostriatal dopaminergic neurons which provide the dopaminergic striatal innervation. Due to the complex chemoarchitectiture of the Central Nervous System (CNS), drug delivery to a very restricted region of the brain is always required. The treatment of Parkinson's disease patients with the dopamine biosynthetic precursor, levodopa in conjunction with a decarboxylase inhibitor [1] has received wide acceptance as an effective approach for the reduction of extrapyramidal symptoms in Parkinson's disease. With the current convential medication levodopa can cause serious adverse effect reactions and its effectiveness decreases with time. However, implanted polymeric devices releasing the appropriate pharmacological agent could restore neurotransmission and lead to functional improvement.

Microparticles for CNS delivery are interesting because even classical well-established small or medium molecular weight drugs intended for the CNS are delivered with difficulties and also new, potentially very active substances present a new challenge in delivery to the CNS because of their inherent chemical properties that present satisfactory delivery by the usual methods[2]. Therefore, several polymer-drug devices have been developed to release drugs to the brain over extended periods of time. These polymeric devices deliver high concentrations of drug to the brain than can be achieved with systemic drug administration while minimizing systemic exposure to the drug, reducing systemic toxicity, increasing absorption and bioavailability and patient compliance to the drug.

In recent years attention has been paid to microparticulate systems of biodegradable polymers for drug delivery. Poly (d,l-lactic acid) (PLA), and poly (d,l-lactide-co-glycolide) (PLAGA) are amorphous, water insoluble, nontoxic, aliphatic polyesters. They are

Biomedical Science and Technology
Edited by Hıncal and Kas, Plenum Press, New York, 1998

biocompatible and biodegradable polymers in vivo because they undergo slow hydrolytic deesterification to CO_2 and H_2O.

The most widely used methods for the preparation of microspheres are solvent evaporation techniques based on o/w or w/o/w emulsions[3]. It has been reported by several groups[3-5] that type and intensity of the dispersion method used to prepare the different emulsions are important parameters affecting particle size, morphology and thus drug loading efficiency and release characteristics of the resulting microspheres.

In the present investigation a 3^2 factorial design has been used to optimize the formulation of biodegradable levodopa microspheres. This approach has the ability to provide the optimal combination of variables at different levels. The two independent variables considered in this study are emulsifying agent concentration and polymer concentration. The dependent variables investigated include $t_{50\%}$; the mean particle size and encapsulation efficiency.

The present study was undertaken to compare different formulations for the preparation of polylactide microspheres loaded with the antiparkinsonian drug, levodopa. Formulations were characterized by particle size, surface morphology, encapsulation efficiency, % yield value and in vitro release properties.

MATERIALS & METHODS

L-PLA (MW= 52000) and D,L-PLA (MW= 33600) polymers were purchased from Sandoz Pharma AG (Basel, Switzerland) whereas PLAGA (MW= 23100) polymer was purchased from Sandoz Pharma AG (Basel, Switzerland). Polyvinyl alcohol (PVA, 99-100 % hydrolyzed, viscosity of 4 % aq.soln. 20 °C, 55-65 cp) was purchased from J.T. Baker Chemical Co. (New Jersey, USA). Levodopa was a generous gift from Merck Sharp Co., (Istanbul, Turkey).

Preparation of Microspheres

Levodopa microspheres were prepared according to the "solvent evaporation" method. Solution of polymers in dichloroethane were dispersed in 100 mL of carboxymethylcellulose (CMC) : sodium oleate (SO) (4:1) or polyvinyl alcohol (PVA): sodium oleate (SO) (4:1) solution whilst vigorously stirring. After removing the organic solvent, the microspheres were centrifuged 3 times at 2000 rpm, washed several times with distilled water, filtered and dried in a dessicator for 24 hrs at room temperature.

Application of Factorial Design

3^2 factorial design experiments were performed to optimize the formulation of biodegradable levodopa-microspheres. The independent variables in the 3^2 factorial design were emulsifying agent concentration (X_1) and polymer concentration (X_2). The effect of the independent factors on the dependent variables, $t_{50\%}$, the mean particle size and encapsulation efficiency was evaluated with the analysis of variance (ANOVA). The multiple regression was applied to dependent variables against independent variables.

Calculation of Percentage Yield Value

The percent yield value of microspheres were determined from the ratio of the amount of products to that of loaded powders.

Table 1. The independent variables and their levels investigated in the optimization of biodegradable levodopa microspheres

Variables	Levels		
Emulsifying Agent concentration * (%), (X1)	0.25	0.50	0.75
Polymer concentration (%), (X2)	6	8	10

* Two different emulsifying agent combinations used were as follow:
1) Carboxymethylcellulose (CMC) : Sodium oleate (SO) (4:1) (w/w) or
2) Polyvinyl alcohol (PVA): Sodium oleate (SO) (4:1) (w/w)

Determination of Encapsulation Efficiency

For each batch, a 10 mg sample was weighed. Washed three times by 1 mL 0.1N HCl. After washing, it is centrifuged at 3500 rpm for 10 min, and the supernatant is assayed at 280 nm spectrophotometrically. Five mL dichloromethane:methanol (4:1) mixture was added to the remaining microspheres followed by centrifugation (3500 rpm) to completely separate the precipitated copolymer. The amount of levodopa in each sample was determined by measuring the absorbance of clear supernatant in a spectrophotometer at 280 nm.

Physicochemical Characterization

Morphological Studies. Scanning electron microscopy was used to examine shape and surface characteristrics of microspheres. Microspheres were mounted on metal stubs with conductive silver paint and then sputtered with a 150 Å thick layer of gold in a BIO-RAD (England) sputter apparatus. The morphological characteristics of the microspheres were examined by Jeol Scanning Electron Microscope (SEM ASID-10) (Japan) in 80 KV.

Particle Size Analysis. Particle size distributions were determined using Coulter-Counter ® (Coulter Multisizer II, Coulter Electronics, England). Samples were prepared by suspending 5 mg of microspheres in 100 mL Isoton® containing 0.01 % w/v Tween 80 for 3 minutes in ultrasonic bath. Particles ranging in size from 1 μm to 30 μm are measurable. Relative number distributions were statistically described as mean and geometric standard deviation.

In vitro Drug Release Studies. The USP XXII rotating basket dissolution apparatus was used at 50 rpm. Fifty milligrams of microspheres were accurately weighed and added to 900 mL of the dissolution medium (0.1 N HCl (pH: 1.2) containing 0.01 % w/v Tween 80) maintaned at 37°C ± 0.5°C. Five millilitre samples were withdrawn at specified time intervals, and were replaced by equivalent volumes of dissolution medium kept at 37°C. The concentration of levodopa was determined by UV spectrophotometry at 280 nm.

RESULTS AND DISCUSSION

The effect of concentration and type of polymer and emulsifying agent on the mean particle size, $t_{50\%}$ and encapsulation efficiency were evaluated using 3^2 factorial design.

Figure 1-6 show the three-dimensional response surface diagrams counter plots for $t_{50\%}$, the mean particle size and encapsulation efficiency of levodopa microspheres, respectively, with variation in 10% polymer concentration and 0.75% emulsifying agent concentration.

The quadratic equations relating mean particle size, encapsulation efficiency and $t_{50\%}$ of microspheres, as a function of polymer concentration and emulsifying agent concentration are:

$$t_{50\%} = 4.51 +7.20x_1-3.47 x_1^2 -1.73 x_2 + 0.158x_2^2 + 0.0250 x_1 x_2 + \varepsilon$$
(Equation 1)

$$\%\text{Encapsulation Efficiency} = -5.22+ 28x_1-10.7x_1^2 +1.83x_2 -0.0417 x_2^2 -0.5 x_1 x_2 + \varepsilon$$
(Equation 2)

$$\text{Mean Particle Size} = -8.38+ 0.407x_1+ 0.720x_1^2+2.58 x_2 -0.126 x_2^2 -0.285x_1 x_2 + \varepsilon$$
(Equation 3)

These equations and Figure 1-6 show that for the preparation of biodegradable levodopa microspheres with high % encapsulation efficiency, selection of polymer concentration as well as emulsifying agent concentration is critical.

A total of 54 formulations were prepared by two different emulsifying agents (PVA:SO; CMC:SO), three different polymers (L-PLA, D,L-PLA and PLAGA) with $t_{50\%}$, % encapsulation efficiency and mean particle size as independent variables. In order to determine the difference between these formulations co-variance analysis is realized. The effectiveness of the independent variables, polymer concentration (x_1) and emulsifying agent concentration (x_2) and their quadratic levels (x_1^2 x_2^2 x_1 x_2) on $t_{50\%}$, % encapsulation efficiency and mean particle size were cleared. After these process, the difference between these two emulsifying agents found statistically significant ($p<0.001$). The difference between the three polymers were also statistically significant ($p<0.01$).

Figure 7 displays the release profile of levodopa loaded biodegradable microspheres under the optimum conditions. Figure 7 also compares the release pattern of microspheres formulated in this study, with those prepared using two different emulsifying agents. In vitro levodopa release studies showed sustained release profile with about 11% - 9% of the total levodopa released in 1 hour followed by an increase to 33% - 32% release in 8 hours for formulations prepared using PVA:SO (4:1) and CMC:SO (4:1), respectively.

Since majority of the surface-adsorbed drug is unavailable for drug targeting, determination of entrapped drug content is more appropriate for the evaluation of a targeted delivery device. Entrapped drug content is generally determined after suitable washings of the carrier[6,7]. Determination of encapsulation efficiency after three washings revealed that the use of 10 per cent w/v of polymer solution and a 0.75% (w/v) emulsifying agent concentration results in maximum drug entrapment. As seen in Table 2, approximately 20.1 and 14.5 per cent of drug can be incorporated into D,L-PLA microspheres prepared using 10 per cent w/v polymer solution and 0.75% CMC:SO (4:1) and PVA:SO (4:1), respectively. Table 2 shows that encapsulation efficiency from the microspheres was different depending on the polymer and emulsifying agent type and concentration. As the concentration of polymer was increased, the encapsulation efficiency increased, too.

In general, increase in emulsifying agent concentration from 0.25% to 0.75% reduced the mean particle size (Table 2). An emulsifying agent concentration of 0.75% (w/v) ensures a good emulsification process and therefore leads to smaller particles. When PVA:SO (4:1) and CMC:SO (4:1) combinations were compared, much smaller particle sizes are obtained with CMC:SO (4:1).

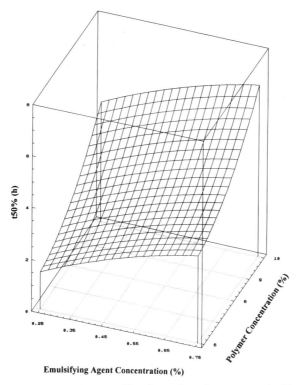

Figure 1. Response surface diagram for $t_{50\%}$ of levodopa microspheres prepared with variation in polymer and emulsifying agent concentration.

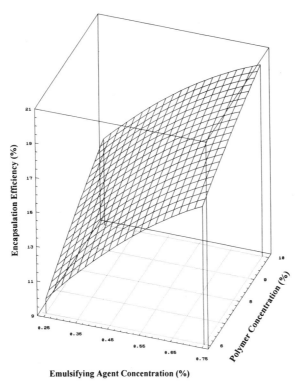

Figure 2. Response surface diagram for encapsulation efficiency of levodopa microspheres prepared with variation in polymer and emulsifying agent concentration.

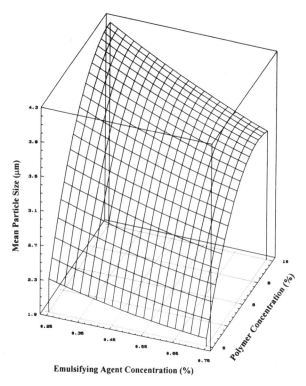

Figure 3. Response surface diagram for the mean particle size of levodopa microspheres prepared with variation in polymer and emulsifying agent concentration.

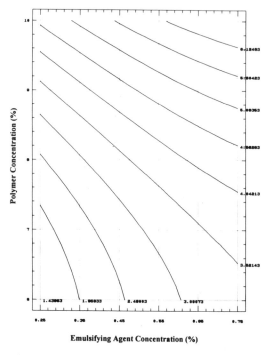

Figure 4. Counter Plots for $t_{50\%}$ of levodopa microspheres prepared with variation in polymer and emulsifying agent concentration.

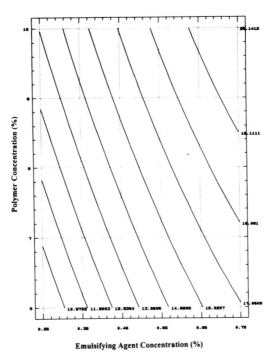

Figure 5. Counter Plots for encapsulation efficiency of levodopa microspheres prepared with variation in polymer and emulsifying agent concentration.

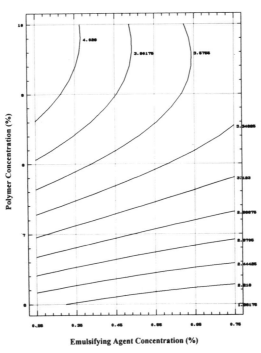

Figure 6. Counter Plots for the mean particle size of levodopa microspheres prepared with variation in polymer and emulsifying agent concentration.

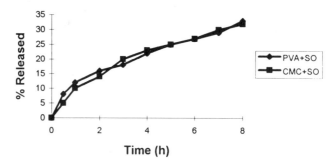

Figure 7. Release profiles of levodopa loaded microspheres for optimum formulations.

As can be seen from the table, the particle size of the microspheres also change ranging from 2.34 to 10.45 μm according to the type and concentration of the polymer, When L-PLA, D,L-PLA and PLAGA polymers are compared, much higher particle sizes were obtained with PLAGA.

Table 2 indicates that there is no statistical difference between percent yield of PVA:SO and CMC:SO formulations (p<0.01). As the amount of emulsifying agent concentrations increased the percent yield has been decreased when both emulsifiers are used.

Table 2. Physicochemical characteristics of microspheres

Polymer	Yield Value* (% ± SD)		Encapsulation Efficiency* (% ± SD)		Particle Size* (μm ± GSD)	
	PVA:SO	CMC:SO	PVA:SO	CMC:SO	PVA:SO	CMC:SO
L-PLA	88± 2.5	86± 1.4	15.0± 2.45	18.9± 1.54	4.95± 1.14	3.74± 0.755
	85± 1.7	82± 2.7	18.6± 1.75	19.2± 1.25	4.45± 2.35	2.80± 0.425
	67± 2.9	76± 3.2	19.4± 1.14	17.4± 0.745	3.75± 1.75	2.34± 0.315
D,L-PLA	82± 3.1	72± 5.4	14.3± 1.075	14.3± 2.15	6.19± 2.14	4.32± 1.15
	78± 3.3	71± 4.7	12.7± 2.14	18.5± 1.75	4.25± 1.45	3.75± 0.875
	73± 2.9	66± 3.7	14.5± 1.85	20.1± 3.45	4.14± 1.27	3.45± 1.65
PLAGA	67± 1.5	66± 1.6	14.3± 0.445	15.6± 2.15	10.45± 3.25	6.77± 1.75
	63± 2.4	62± 2.3	11.2± 1.12	15.9± 1.15	9.15± 2.15	6.45± 0.545
	59± 3.7	61± 1.9	14.7± 1.47	18.4± 1.37	8.13± 1.55	5.51± 1.85

*(n=6)

A representative photomicrograph of biodegradable levodopa microspheres prepared by solvent evaporation method using 10 per cent w/v of polymer solution and 0.5 per cent w/v of emulsifying agent (CMC:SO (4:1)) is displayed in Figure 8. The batches prepared at other levels of different variables also resulted in the formation of discrete spherical particles.

CONCLUSIONS

Solvent-evaporation method and biodegradable polymers (L-PLA, PLAGA and D,L-PLA) are effective in preparing microspheres containing antiparkinsonian drugs. This study illustrates the use of factorial design in the optimized biodegradable levodopa microspheres and involves investigation of various levels of emulsifying agent and polymer concentration for the development of microspheres with high encapsulation efficiency and T50%.

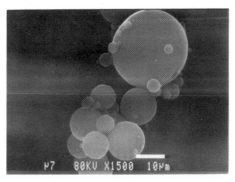

Figure 8. Scanning Electron Microphotograph of Levodopa loaded D,L-PLA microspheres.

Interactions between emulsifying agent and polymer concentrations has been evaluated. Use of high polymer and emulsifying agent concentrations are required for efficient entrapment and small mean particle size. The results obtained by using 10% w/v polymer concentration, which is the highest concentration realized, supported the above stated thesis. A significantly sustained release of levodopa from the microspheres is achieved.

The optimal formulations, having high drug entrapment and appropriate small particle size will be used in the in vivo studies. Further studies are in progress to determine the efficacy of levodopa microspheres in combination with a decarboxylase inhibitor in animal experiments and to evaluate the feasibility of microspheres.

Acknowledgements

This work was supported by a grant from the Hacettepe University, Research Fund (9401013015).

REFERENCES

1. A. Barbeau, 1974, The clinical physiology of side effects in long-term l-dopa therapy, *Adv. Neurol.* **5**:347-356

2. WM. Pardridge, 1996 Brain drug delivery and blood-brain barrier transport. *Drug Delivery* **3**: 99-115.

3. M. Iwata and J.W. McGinity, 1992, Preparation of multiphase microspheres of poly(d,l-lactic-co-glycolic acid) containing a w/o emulsion by a multiple emulsion solvent evaporeation technique, *J. Microencap.* **9**:201-214

4. R. Alex and R. Bodmeier, 1990, Encapsulation of water-soluble drugs by a modified solvent evaporation method. I. Effect of process and formation variables on drug entrapment, *J. Microencap.* **7**:347-355

5. B. Arıca, H.S. Kaş, M. Orman, H.H. Çelik and A.A. Hıncal, 1996, Proceed. 8th Inter. Pharm. Technol. Symp. (IPTS-96), Sept. 9-11, pp.167

6. P. K. Gupta, C. T. Hung and D.G. Perrier, 1986a, Albumin microspheres I: release characteristics of adriamycin. *Int. J. Pharm.* **33**:137-146

7. P. K. Gupta, C. T. Hung and D.G. Perrier, 1986a, Albumin microspheres II: Effect of stabilization temperature on the release of adriamycin. *Int. J. Pharm.* **33**:147-153

SURFACE-MODIFIED PHOSPHOLIPID-STABILIZED EMULSIONS AS

TARGETED SYSTEMS FOR INHIBITION OF METASTATIC CANCER

Michael J. Groves [1] and Xiaoyan Gao[2]

[1]Institute for Tuberculosis Research
College of Pharmacy
University of Illinois at Chicago
Room 2014 SEL (M/C 964)
950 South Halsted Street
Chicago, IL 60607-7019
(USA)
[2]Present Address: Abbott Laboratories Inc.
Dept. 97D Building AP4
100 Abbott Park Road
Abbott Park, IL 60064-3500

INTRODUCTION

It is now recognized that deaths from cancer are most frequently due to metastic development of the initial disease site, irrespective of the treatment or surgical extirpation of the primary tumor following diagnosis. The significant role played by fibronectin in the promotion and support of tumor cell migration from the primary cancer site to peripheral metastatic sites is well documented.[1-6] Humphries[7] pointed out that interference with the function of fibronection would probably inhibit the metastatic process and these authors identified a short peptidal fragment, GRGDS, that functioned in this manner. Unfortunately, the fragment was not stable and has a very short biological half life, requiring frequent administration of large quantities of the peptide.[8] Later[4] Humphries reported that both polymeric and cyclic peptides were also potent inhibitors of integrin-dependent adhesion and experimental metastasis models. Braatz et al.[9] have also shown that a peptide containing the essential RGD peptide sequence retained its biological activity with an extended circulatory half-life when covalently bound to an isocyanate-containing polyurethane polymer.

Gelatin binds fibronectin with high affinity.[10-15] By specifically binding to fibronectin, gelatin could interfere with the binding of cells to fibronectin enriched-surfaces.[16] However, since gelatin is a complex mixture of macromolecules with various range molecular weights, combined with the observation of severe side effects following infusion of gelatin-based plasma substitutes in man,[17-23] the use of peptides that contained the active binding domains would be preferred. Two previously unknown fibronectin-binding peptides have been isolated from trypsinized gelatin.[24]

These peptides were sequenced as:

Peptide I: Thr-Leu-Gln-Pro-Val-Tyr-Glu-Tyr-Met-Val-Gly-Val

Peptide II: Thr-Gly-Leu-Pro-Val-Gly-Val-Gly-Tyr-Val-Val-Thr-Val-Leu-Thr

Although these peptides had higher specific binding activity to human fibronectin than the gelatin sample from which they had been isolated, it seemed likely that the biological activity of these materials would be enhanced by coupling to a suitable adjuvant. In this manner not only would the system be more stable and capable of passively interfering with fibronectin-surfaces during the metastatic process but it should also be capable of targeting the surface with a vehicle which could contain an appropriate antineoplastic agent, thereby not only enhancing the specific activity of the drug but reducing the required dose and associated toxicities.

Accordingly, we have adapted methodologies reported for covalently bonding peptides to phospholipids in liposomes to phospholipid-stabilized emulsions. Emulsion systems are attractive as drug delivery systems for hydrophobic drugs, including potential use as sustained-release carriers, targeting and vaccine adjuvants.[25,26] Unlike liposomes, biocompatible emulsions are readily prepared on an industrial scale and are relatively stable for extended periods of time.[27] We utilized a methodology reported for the most part by Endoh et al.[28,29] in which the peptides are covalent bound through their N- or C-termini to phosphatidylethanolamine (PE), naturally found as a component of commercial phospholipid-stabilized emulsions such as 'Intralipid' (Pharmacia-Upjohn). What was surprising was the observation that this reaction could be carried out in situ on the intact sterilized emulsion, without having to make the emulsion by assembly of suitably modified components followed by disruption by homogenization and subsequent heat-sterilization.

Covalently coupled to an emulsion vehicle ('Intralipid'), these two peptides were anticipated to function as ligands for targeted delivery systems with enhanced activity and longer half-life.[24] Whether the coupling process would destroy the intrinsic binding properties of these peptides and which coupled formulation preserved the highest binding affinity were questions that needed to be answered.

Selected formulations were further studied for inhibition of fibronectin-mediated cell spreading and targeting activity to tumor cells. The stability of the peptide-coupled 'Intralipid'-based emulsions in respect to cleavage of the coupled peptide and particle size changes was also determined.

MATERIALS AND METHODS

Intralipid® 20% i.v. Fat Emulsion (Lot NR57327 B, Pharmacia-Upjohn, Clayton, NC) with a geometric mean diameter of 310 nm (σg = 1.15) (measured with Malvern Zetasizer III, Malvern Instruments Ltd., UK). L-α-phosphatidylcholine (plant) (PC), L-α-phosphatidylethanolamine (egg) (PE) were purchased from Avanti Polar Lipids, Inc. (Alabaster, AL). Thin-layer chromatography plates: K6 Silica Gel 60 Å, 20 x 20 cm were from Whatman Inc. (Hillsboro, OR). Silica Gel HF-254 was from Brinkmann Instruments, (Des Plaines, IL). Tolylene-2,4-diisocyanate (TDIC), 1-ethyl-3-(3-dimethylaminopropyl) carbodiimide (EDIC), molybdenum and molybdenum trioxide were all from Sigma Chemical Co. (St. Louis, MO). The fibronectin-binding peptides, Peptide I, Peptide II and fluorescein-labeled Peptide I at its N- and C-termini were synthesized to a purity of better than 98.5% by Genosys Biotechnologies (Woodlands, TX). Bio-Gel A-1.5m Gel was from Bio-Rad (Hercules, CA). Modified Lowry Protein Assay Reagent was from Pierce (Rockford, IL).

Coupling of N-termini of the Peptides to the 'Intralipid'

To 1 mL of the 'Intralipid' was added 20 μL of a 2% solution of TDIC in p-dioxane. The reaction mixture was incubated at 17°C for 2 hr with gentle shaking. Polymerized

material derived from the TDIC was removed from the mixture by centrifugation (Sorvall®
RC-5B Refrigerated Superspeed Centrifuge, Du Pont Instrument, New Town, CT) at 400
× g for 5 min. The resulting supernatant was then mixed with 1 mL of the peptide solution
(2 mg/mL) and incubated at 37°C for 2 hours to produce peptide-coupled 'Intralipid'. The
peptide-coupled droplets were separated from noncovalently bound peptide and free peptide
by adding 8 mL of 0.9% NaCl containing 0.02% EDTA to the coupled mixture with gentle
agitation, followed by chromatography on a Bio-Gel A-1.5m column (1.5 × 30 cm) with
0.9% NaCl as the elution buffer.

Coupling of C-termini of the Peptides to the 'Intralipid'

To a mixture of the 'Intralipid' (1 mL) and solution of the peptide (1 mL, 2 mg/mL),
was added 0.2 mL of a 20% solution of EDCI. After adjusting the pH to 4.7 with 0.3 N
HCl and standing for 2 hours at 24°C, this mixture was fractionated to give covalently
coupled peptide-emulsion droplets by the separation procedure described above.

Effect of the Coupling Processes on the Size Distribution of the Coupled 'Intralipid'

The size distributions of the emulsion systems before and after the coupling processes
were measured with the Malvern Zetasizer III (Malvern Instruments Ltd., UK). Particle size
number distributions (% undersize) were analyzed using a log-probability plot to obtain the
geometric mean number diameter and geometric standard deviation for each formulation,
Table 1.

TLC Characterization of the PE Alone and PE in 'Intralipid' Before and After the Coupling Processes

The coupled PE and phospholipids from the Peptide I-coupled emulsion system were
extracted and assayed by a method established in this study.[24]

Extraction of the Lipophilic Components From the Peptide I-Coupled Emulsion

Peptide I-coupled emulsion was centrifuged at 15,000 rpm for 5 minutes. The
separated oil- and water-rich phases were each added to 4 mL of chloroform/methanol (2:1,
v/v), 5 mL of chloroform and finally 2 mL of water. After each addition, the mixtures
were shaken vigorously for 1 minute. The extraction mixtures were added to Squibb Pear-
shaped Separatory Funnels and allowed to stand until clear separation had been achieved.
Lipophilic components distributed in the lower chloroform phase were collected and
vacuum-dried.

Separation of Phospholipids and Their Derivatives From the Lipophilic Components

The chloroform extract was redissolved in 1.5 mL of chloroform and loaded onto a
Silica Gel HF-254 column (9 × 1 cm). The column was first washed with 4 mL of
chloroform to remove the oil and then with methanol at a flow rate of 0.5 mL/min.
Phospholipid-rich fractions were collected and vacuum-dried.

Thin-Layer Chromatography (TLC) of the Phospholipid Extracts

Standard PC and PE solutions in chloroform and the dried phospholipid extracts
dissolved in chloroform/methanol (1:1, v/v) were quantitatively applied to a TLC plate.
Spots were developed with chloroform: methanol: 50% ammonium hydroxide (65:25:8) and

colored with the molybdate spray agent due to Ellingson and Lands.[30] The PE spots in the phospholipid extracts were identified by their R_f values compared to those of standards.

Validation of the Couplings

Peptide I samples, labeled with a fluorescein probe (Genosys Biotechnologies) at either end, were used so that the coupling of the peptide to the PE in the emulsion could be visualized under a fluorescent microscope.

Preparation of Fluorescein-Labeled Peptide I (F-Peptide I) Coupled to Emulsion at Either End

The coupling of the F-Peptide I to the emulsion at either end followed the same procedures as described above, except that Peptide I with fluorescein labeled at its N-terminus was used to couple to the emulsion at its C-terminus, and *vice versa*.

Fluorescent Microscopy (Department of Anatomy and Cell Biology, UIC)

A Leitz Orthomat microscope filled with an E2-E3 fluorescence cube for fluorescence microscopy was used. The filter system was adjusted to give an excitation wavelength of 494 nm and emission of 520 nm.

Coupling Kinetics of Peptide I onto the Emulsion

Various amounts of Peptide I were coupled to the emulsion system at either the N- or C-termini in the presence or absence of the coupling agents (i.e. TDIC and EDCI). Peptide I that was not coupled to the emulsion was separated from the coupled system by the same procedure described above. In the absence of coupling agents, the emulsion was simply incubated with Peptide I at room temperature for 2 hours and then chromatographed on the Bio-Gel A-1.5m column with or without the addition of 0.02% EDTA. The amount of Peptide I remaining coupled or adsorbed (no coupling agents) to the emulsion was collected after three cycles of trichloracetic acid precipitation and measured by the Modified Lowry Protein Assay Reagent, using Peptide I alone as a standard.

Determination of the Binding of the Peptides Coupled 'Intralipid' to Fibronectin

The binding affinity of the peptides and their coupled 'Intralipid' were determined by an ELISA method. Briefly, microtiter wells (96-well vinyl EIA/RIA plates, Gibco, Grand Island, NY) were first coated with peptide or its coupled emulsion prepared as described and incubated at 4°C for two days. After the removal of the supernatants, the wells were washed with PBS-T [0.01 M phosphate buffer saline pH 7.4 (PBS) supplemented with 0.05% Tween-20 and 0.02% sodium azide]. The wells were then blocked with blocking buffer [0.25% bovine serum albumin (BSA)] for 1 hr at 37°C. After a serial dilution of fibronectin was added and incubated at 37°C for 2 hr, anti-fibronectin antiserum (Rabbit, Calbiochem, San Diego, CA) in PBS-T (1:1000) was added and incubated at 37°C for 1.5 hr. The wells were then further incubated with the Goat-anti-Rabbit Ig G (H+L) alkaline phosphatase conjugate (Bio-Rad, Hercules, CA) diluted 1000-fold with conjugate buffer (0.05 M Tris + 1% BSA + 0.02% sodium azide, pH 8.0) at 37°C for 30 min. The plates were incubated with 50 μL of the substrate buffer (0.05 M glycine + 1.5 mM magnesium chloride, pH 10.5) and substrate (p-nitrophenyl phosphate disodium, Sigma, St. Louis, MO, 4 mg/mL) for 15 min at 37°C. At the end of the incubation period, the reaction was terminated by the addition of 1 M NaOH and the absorbance at 410 nm measured using a

Dynatech plate reader (MR 300, Dynatech Laboratories, Chantilly, VA). The wells with coating buffer served as controls.

The binding data were analyzed by a nonlinear regression process using GraphPad Prism™, version 2.01 (San Diego, CA). Data were fitted to both one-site binding and two-site binding models and compared. The best fit model was selected by the program and the binding constants calculated according to the models:

One site binding: $A = A_{max} \bullet [L]/\{K_d+[L]\}$ Eq (1)

Two site binding: $A = A_{max1} \bullet [L]/\{K_{d1}+[L]\} + B_{max2} \bullet [L]/\{K_{d2}+[L]\}$ Eq (2)

where, A is the absorbance at 410 nm, A_{max} is the maximum absorbance at saturation; [L] is the molar concentration of the free fibronectin, and K_d is the dissociation constant.

Interaction of Fibronectin-Binding Peptide Coupled 'Intralipid' with S180 Sarcoma Cells

Murine sarcoma (S180) cells express fibronectin on their surface[31], and by binding to the fibronectin, gelatin microparticles also inhibited S180 adherence to a polystyrene substrate.[15] By adapting the same methodology, the targeting activity to fibronectin-bearing tumor cells of Peptide I coupled to the emulsion at its N-terminus (PIN-E) could be measured.

Murine sarcoma (S180) cells were obtained from the American Type Culture Collection (Rockville, MD). The cells were maintained in CMEM-E (Eagle's minimum essential medium with non-essential amino acids, Earle's basal salts, 5% calf serum, 100 units/mL penicillin and 100 μg/mL streptomycin) in an atmosphere of 5% CO_2, 95% relative humidity at 37°C. Viable S180 cells, 10^6 per well, in MEM-E (Eagle's medium without serum) (0.9 mL) plus 0.1 mL of different concentrations of PIN-E (concentration at zero as control) were distributed among the center eight wells of cluster plates. After 5 hr incubation at 37°C, nonadherent cells (found to be associated with PIN-E) were resuspended by gentle aspiration three times with phosphate-buffered saline (PBS) from a Pasteur pipette and removed. Adherent cells (most of which were not associated with PIN-E) were harvested and counted in a hemocytometer. Inhibition of the targeting of PIN-E to S180 cells by fibronectin was carried out by pre-incubating PIN-E (equivalent to 31.08 μg/well Peptide I) with a series of fibronectin concentrations and the rest of the experiment was carried out as above.

Inhibition of Fibronectin-Mediated Cell Spreading

a. Inhibition of Fibronectin-Mediated Cell Spreading by PIN-E

Fibronectin-mediated cell spreading of baby hamster kidney (BHK) cells was assayed as described by Grinnell[32] and Yamada and Kennedy.[33] BHK cells (BHK-21) were obtained from the American Type Culture Collection (Rockville, MD) and maintained in minimum essential medium (Eagle) with non-essential amino acids, 90% Earle's BSS, 10% fetal bovine serum. Tissue culture clusters (24-well, Coster, Cambridge, MA) were pre-incubated with 3 μg/mL of fibronectin in an adhesion medium (150 mM NaCl, 3 mM KCl, 1 mM $CaCl_2$, 0.5 mM $MgCl_2$, 6 mM Na_2HPO_4, 1 mM KH_2PO_4, pH 7.3) at room temperature for 60 min and the nonspecific adsorption sites were blocked with 10 mg/mL heat-denatured (80°C for 30 min) BSA for 30 min. BHK cells were trypsinized, washed three times with PBS and incubated in adhesion medium with or without added various concentrations of PIN-E for 45 min at 37°C. After the attachment period, cells were fixed with 2.5% glutaraldehyde in PBS for 1 hr, and cell spreading measured as described by Yamada and Kennedy.[33] Controls for background spreading were carried out on the wells that had not been coated with fibronectin; controls for the possible nonspecific effect of protein present during the cell spreading assay included incubation with various concentrations of BSA.

b. Competitivity of the Inhibition Effect of PIN-E by Fibronectin

The competitivity of the inhibition by fibronectin was studied by pre-incubation of 24-well tissue culture clusters with various concentrations of fibronectin in the adhesion medium for 60 min at room temperature. Non-specific sites blocking with BSA and incubation of BHK cells with or without PIN-E were carried out as described above.

c. Reversibility of the Inhibition Effect of PIN-E

After the inhibition of BHK cells spreading by PIN-E, the cells were washed three times with adhesion medium and incubated in the adhesion medium only for another 45 min at 37°C. The cell spreading was quantitated again under the same conditions.

Cleavage of Peptide I of PIN-E upon Storage, in Serum

Cleavage upon storage at 4°C and 37°C was tested by diluting PIN-E in an equal volume of 0.9% NaCl and kept in vials. Mixtures with rabbit serum (Gibco, Grand Island, NY) were prepared in the same manner and stored at 37°C. Aliquots were removed at different times and cleaved Peptide I separated from coupled material by using Ultra Spin Centrifuge Filters-Cellulose (5 kDa MWCO, Alltech, Deerfield, IL) with two consecutive washes of 0.9% NaCl. The filtrate washes were pooled for the peptide assay. For PIN-E incubated with rabbit serum, the Peptide I remaining coupled was separated from cleaved material by chromatography on Bio-Gel A-1.5m (Bio-Rad, Hercules, CA). The amount of Peptide I remaining coupled to 'Intralipid' was collected after three cycles of trichloroacetic acid precipitation.

The Peptide I was measured with a Modified Lowry Protein Assay Reagent (Pierce, Rockford, IL), using Peptide I as standard.

Changes in Size of PIN-E upon Storage, in Serum

Change in size distribution of PIN-E upon storage at 4°C or room temperature and in rabbit serum at 37°C were analyzed with respect to the geometric mean diameter (Malvern Zetasizer III, Malvern Instruments Ltd., UK).

RESULTS AND DISCUSSION

The presence of phosphatidylethanolamine (PE) in the phospholipid mixture used to stabilize the emulsion makes it ideal for coupling with protein or peptide drugs. However, whether those established coupling methods could be adapted to an emulsion system in situ was not known. The coupling reaction was therefore first checked by TLC. PE spots after coupled to N- or C-terminus of Peptide I, showed significant decreases in both size and color density whereas new spots with significant size and density differences as apparent replacements for PE emerged.[24] These suggested that the available intact PE had reacted, with new products forming from the coupling reaction.

In addition, the coupling reaction in our system was visualized by using fluorescein-labeled Peptide I. Under the fluorescent microscope, concentrated fluorescence associated with the emulsion particles could be seen.[24] These observations provide further direct evidence that coupling of Peptide I to 'Intralipid' in situ using cross-linking agents was feasible.

Peptide I coupled significantly to the 'Intralipid' emulsion with the cross-linking agents, TDIC or EDCI. Non-specific binding of the peptide to the 'Intralipid' was also observed in the absence of the cross-linking agents but the amount of Peptide I coupled directly onto

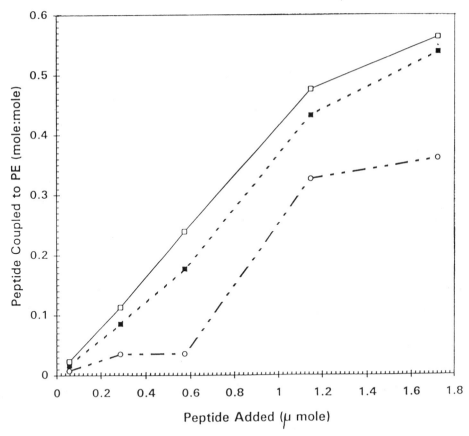

Figure 1. Coupling kinetics of Peptide I onto 'Intralipid' with various amounts of Peptide I in the presence or absence of the cross-linking agents.

the 'Intralipid' was related to the amount added in the reaction. As shown in Figure 1, the amount of Peptide I coupled to the 'Intralipid' increased almost linearly, to a saturation limit at ~ 0.43 mole/mole PE (N-terminus) and 0.48 mole/mole PE (C-terminus).

Table 1. Size parameters of 'Intralipid' before and after coupling to fibronectin-binding peptides at either termini

Size parameters	'Intralipid'	PIN-E	PIC-E	PIIN-E	PIIC-E
d_g (nm)	310	320	750	400	690
σ_g	1.15	1.23	1.47	1.25	1.44

Note: PIN-E: Peptide I coupled to 'Intralipid' at its N-terminus.
         PIC-E: Peptide I coupled to 'Intralipid' at its C-terminus.
         PIIN-E: Peptide II coupled to 'Intralipid' at its N-terminus.
         PIIC-E: Peptide II coupled to 'Intralipid' at its C-terminus.
         d_g: Geometric mean diameter on a number basis.
         σ_g: Geometric standard deviation on a number basis.

Particle diameters of the emulsion systems are shown in Table 1. It appeared that the geometric number mean diameters of the coupled particles significantly changed when compared to those of the original 'Intralipid' system. An exception was observed for Peptide I coupled to 'Intralipid' droplets at its N-terminus, PIN-E, in which only minor increases in droplet size were observed, Table 1.

Peptide I, Peptide II and their coupled formulations all fitted a two-site binding model. Since we were mainly interested in the binding site with the higher binding affinity, only the Ka_1 is discussed here. Peptide I and Peptide II maintained their binding affinities to fibronectin even after coupling to the emulsion through their N-termini (Table 2) although Peptide I lost its binding affinity significantly after coupling through its C-terminus. Comparison of Ka_1 values for PIN-E and PIIN-E did not demonstrate significant differences, suggesting that PIN-E and PIIN-E had similar binding affinities to fibronectin. Since PIN-E had a higher binding affinity and the particle size distribution was relatively unchanged after coupling, this system was selected for further study.

At a fibronectin concentration of 250 μg/well, the inhibition of the adherent S180 cells by PIN-E was completely abolished. This result also indicates that the interference of S180 cell adherence by PIN-E was entirely due to the interaction between Peptide I and fibronectin on the cell surfaces. As shown in Figs. 4 and 5, BHK cells spread completely on tissue culture substrates precoated with 3 μg/mL human plasma fibronectin, but hardly spread on the substrate surfaces not precoated with fibronectin. This fibronectin-mediated spreading was progressively inhibited by increased concentrations of PIN-E added to the adhesion medium. Inhibition appeared maximal at an equivalent Peptide I level of 932 μg/well. Although spreading was severely inhibited, some cells appeared to display abortive spreading, with an increased phase density of cytoplasm according to phase-contrast microscopy, but with poor elaboration of peripheral lamellae.[24] BSA at various concentrations as controls for nonspecific effects of added protein did not affect fibronectin-mediated spreading of these cells.

The inhibition of BHK cells spreading by PIN-E appeared to be competitive, since the inhibitory effect was diminished in a dose-dependent fashion as the amount of fibronectin preadsorbed onto the substrate was increased. As shown in Fig. 5, there was an $\sim 27\%$ inhibition of spreading by the PIN-E (equivalent to 466 μg/well Peptide I) at a low concentration of fibronectin in the adsorption step, but this had decreased to $< 15\%$ inhibition at the highest concentration. Removal of the PIN-E by washing resulted in a rapid re-initiation of spreading approaching to 100%, indicating that the inhibition of BHK cells spreading by PIN-E was not due to its cytotoxicity, but rather could be totally attributed to interaction with fibronectin, Fig. 6.

Table 2. Ka_1 for gelatin, isolated peptides PI and PII and their emulsion-coupled systems interacting with fibronectin

	Ka_1 $M^{-1} \times 10^{-9}$
gelatin	0.19
PI	1.48
PIN-E	1.24
PIC-E	0.03
PII	1.30
PIIN-E	0.23
PIIC-E	0.73

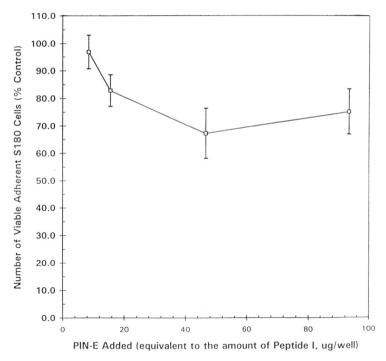

Figure 2. The effect of increasing doses of PIN-E at 37°C on the association of murine S180 sarcoma cells for the plastic walls of the wells in the cluster plate. As the concentration of PIN-E increased, the number of adherent cells decreased. Bar = standard error (n = 8).

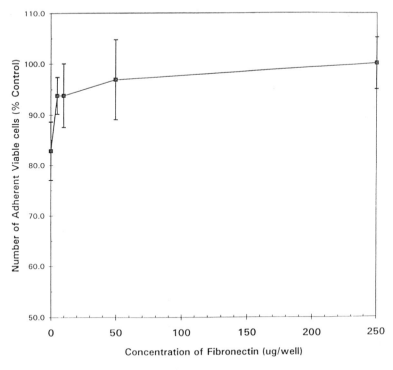

Figure 3. The effect of fibronectin on the interaction between PIN-E and S180 sarcoma cells _in_ _vitro_ at 37°C. Bar = standard error (n = 4).

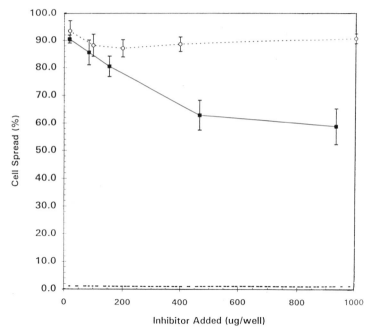

Figure 4. Inhibition of fibronectin-mediated spreading of BHK cells by increased concentrations of added PIN-E. Cells were placed into dishes precoated with 3 μg/mL plasma fibronectin and incubated in the presence of the indicated concentrations of added PIN-E. The background level of cell spreading in dishes not coated with fibronectin and BHK spreading in the presence of various BSA concentrations were as indicated. Each point represents data from approximately 400 cells; bar indicates standard error for 10 random microscope fields.

-■- :PIN-E (equivalent to the amount of Peptide I) ⋯o⋯ :BSA
⋯⋯⋯ :Background spreading on wells not coated with fibronectin

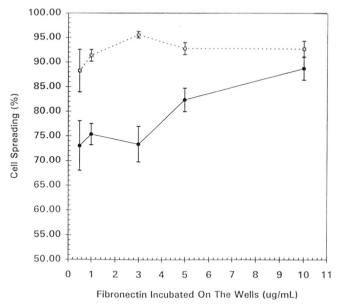

Figure 5. Competitive nature of the inhibition by PIN-E. The graph indicates the percentage of total BHK cells with a spread morphology coated with the indicated concentrations of plasma fibronectin in the absence or presence of PIN-E (equivalent to 466 μg/well Peptide I). Points indicate mean ± standard error of fourteen random fields (400 cells each). --⊠--Control —— Std. Error (control) ——●—— PIN-E —— Std. Error (PIN-E)

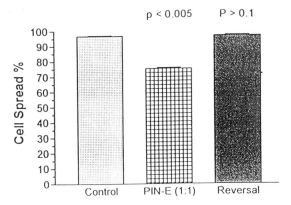

Figure 6. Reversibility of the inhibition of PIN-E. Percentage of BHK cells with a flattened, spread morphology on a control fibronectin substrate after precoating with 3 μg/mL plasma fibronectin (A) or the same substrate plus PIN-E (equivalent to 466 μg/well Peptide I) in adhesion medium (B), or after the same treatment with PIN-E for 45 min, followed by three washes and a further incubation for 45 min in the absence of the inhibitor (C). Error bars indicate standard error, n = 30; 600 cells for each point.

From a pharmaceutical perspective, for which a relatively stable system is required, PIN-E did not show any significant cleavage of Peptide I when stored at 4°C for 22 days, while, at room temperature, the cleavage began to increase significantly after the sixth day until, on the 22nd day, only 55% Peptide I remaining coupled, Fig. 7. It is apparent that storage at 4°C is more appropriate. In order to predict the stability of PIN-E, data at 4°C was treated by zero-, first- or second order kinetics and the significance of linear relationships between the percentage remaining coupled and time tested using an F-test. It appeared that although storage at 4°C did not induce any significant cleavage of Peptide I from PIN-E, subsequent data variation might simply be attributed to experimental error.

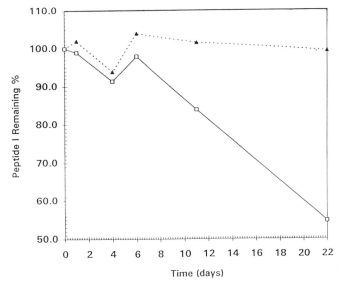

Figure 7. Peptide I remaining coupled (%) upon storage at 4°C and room temperature.
--- ▲ --- 4°C —□— Room temperature

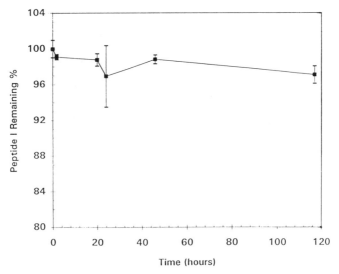

Figure 8. Peptide I remaining coupled upon incubation with rabbit serum at 37°C. All values represent mean and standard error (n = 3).

PIN-E showed some immediate cleavage of Peptide I when incubated with rabbit serum for ~ 2 hr at 37°C. However, continuous 117 hr incubation only resulted in a total of 3% cleavage, indicating that PIN-E would remain in its effective form in the blood circulation before reaching its targets, Fig. 8.

Particle size analysis of PIN-E stored at 4°C demonstrated that the droplet diameter remained unaltered for at least one month and showed no trend to increase when stored at room temperature, the size of PIN-E emulsion droplets increased significantly over 20 days, Fig. 9. Incubation of PIN-E in rabbit serum did not induce any significant change of size for the first day, Fig. 10.

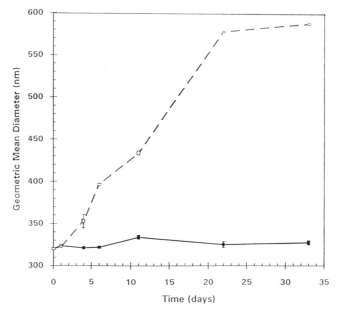

Figure 9. Changes in the size of PIN-E upon storage at 4°C and room temperature. Error bars indicate standard error, n = 10.

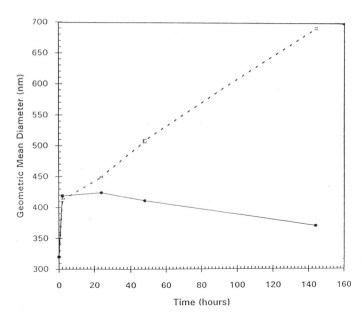

Figure 10. Changes in droplet size of PIN-E upon incubation with rabbit serum at 37°C. Points indicate mean ± standard error of 10 measurements.

All these results suggested that PIN-E is stable at 4°C and, after injected in blood, its size and coupled peptide should be maintained long enough for it to reach its intended targets. The effects due to other factors in the circulation remain to be demonstrated and in vivo testing remains to be carried out.

The fibronectin-binding peptide vectored emulsion, by interaction with fibronectin, inhibits fibronectin-mediated cell spreading in the experimental model and again by its high binding affinity to fibronectin, should target to a fibronectin-coated tumor cell. This strongly suggests that this fibronectin-binding peptide vectored drug delivery system could become a promising alternative device delivery of antineoplastics in cancer treatment.

REFERENCES

1. S.K. Akiyama, K. Olden, and K.M. Yamada, 1995, Fibronectin and integrins in invasion and metastasis, *Cancer and Metastasis Review* **14**(3):173-189.
2. S.K. Akiyama, K. Nagata, and K.M. Yamada, 1990, Cell surface receptors for extracellular matrix components, *Biochim. Biophys. Acta.* **1031**:91-110.
3. M.J. Humphries, 1990, The molecular basis and specificity of integrin-ligand interactions, *J. Cell Sci.* **97**:585-592.
4. M.J. Humphries, 1993, Fibronectin and cancer: Rationales for the use of antiadhesive in cancer treatment, *Cancer Biology* **4**:293-299.
5. R.O. Hynes, 1992, Integrins: Versatility, modulation and signaling in cell adhesion, *Cell* **69**:11-25.
6. K.M. Yamada, 1991, Adhesive recognition sequences, *J. Biol. Chem.* **266**:12809-12812.
7. M.J. Humphries, K. Olden, and K.M. Yamada, 1986, A synthetic peptide from fibronectin inhibitors experimental metastasis of murine melanoma cells, *Science* **233**:467-470.
8. M.J. Humphries, K.M. Yamada, and K. Olden, 1988, Investigation of the biological effects of anti-cell adhesive synthetic peptides that inhibit experimental metastasis of B16-F10 melanoma cells, *J. Clin. Invest.* **81**:782-790.
9. J.A. Braatz, Y. Yasuda, K. Olden, K.M. Yamada, and A.H. Heifetz, 1993, Functional peptide-polyurethane conjugates with extended circulatory half lives, *Bioconjug. Chem.* **4**(4):262-267.
10. D.F. Mosher, 1980, Fibronectin, *Prog. Hemostasis Thromb.* **5**:111-151.
11. H. Forastieri, and K.C. Ingham, 1983, Fluid-phase interaction between human plasma fibronectin and gelatin determined by fluorescence polarization assay, *Arch. Biochem. Biophys.* **227**:358-366.

12. K.C. Ingham, S.A. Brew, and B.S. Isaacs, 1988, Interaction of fibronectin and its gelatin-binding domains with fluorescent-labeled chains of type I collagen, *J. Biol. Chem.* **263**(10):4624-4628.

13. A. Garcia-Pardo, and L.I. Gold, 1993, Further characterization of the binding of fibronectin to gelatin reveals the presence of different binding interactions, *Arch. Biochem. Biophys.* **304**(1):181-188.

14. K. Nakamura, S. Kashiwagi, and K. Takeo, 1992, Characterization of the interaction between human plasma fibronectin and collagen by means of affinity electrophoresis, *J. Chromat.* **597**:351-356.

15. Y. Lou, W.P. Olson, X.X. Tian, M.E. Klegerman, and M.J. Groves, 1995, Interaction between fibronectin-bearing surfaces and Bacillus Calmette Guérin (BCG) or gelatin microparticles, *J. Pharm. Pharmacol.* **47**:177-181.

16. K. Nagata, M.J. Humphries, K. Olden, and K.M. Yamada, 1985, Collagen can modulate cell interactions with fibronectin, *J. Cell Biol.* **101**:386-394.

17. D.L. Heene, D. Zekorn, and H.G. Lasch, 1968, Gelatin plasma volume expanders: Chemistry, biological activities and clinical experiences, *Proc. 11th Congr. Int. Soc. Blood Transf.*, Sydney 1966, Bibl. No. 29, Part 3, pp. 907-913, Karger, Basel, New York.

18. H.H. Schöne, 1960, Chemistry and physicochemical characterization of gelatin plasma substitutes, Modified Gelatin as Plasma Substitutes, *Biol. Haemat.*, No. 33, pp. 78-90, Karger, Basel, New York.

19. D. Zekorn, 1969, Intravascular retention, dispersal, excretion and breakdown of gelatin plasma substitutes, Modified Gelatin as Plasma Substitutes, *Biol. Haemat.*, No. 33, pp. 131-140, Karger, Basel, New York.

20. G. Brodin, F. Hesselvik, and H. von Schenck, 1984, Decrease of plasma fibronectin concentration following infusion of a gelatin-based plasma substitute in man, *Scand. J. Clin. Lab. Invest.* **44**:529-533.

21. J.M. Saddler, and P.J. Horsey, 1987, The new generation gelatins: A review of their history, manufacture and properties, *Anesthesia* **42**:998-1004.

22. J.M. Vedrinne, J.P. Hoen, D. Bussery, C. Veyssere, M. Richard, and J. Motin, 1991, Plasma fibronectin and complement following infusion of colloidal solutions after spinal anaesthesia, *Intensive Care Med.* **17**:83-86.

23. F.A. Blumenstock, P.L. Celle, A. Herrmannsdoerfer, C. Giunta, F.L. Minnear, E. Cho, and T.M. Saba, 1993, Hepatic removal of ^{125}I-DLT gelatin after burn injury: A model of soluble collagenous debris that interacts with plasma fibronectin, *J. Leukocyte Biol.* **54**:56-64.

24. X. Gao, 1996, Peptides derived from gelatin as vectors for potential cancer treatment, *Ph.D. Thesis*, University of Illinois at Chicago.

25. Y. Yamaguchi, and Y. Mizushima, 1994, Lipid microspheres for drug delivery from the pharmaceutical viewpoint, *Critical Reviews in Therapeutic Drug Carrier Systems* **11**(4): 215-229.

26. D.M. Lidgate, and N.E. Byars, 1995, Development of an emulsion-based muramyl dipeptide adjuvant formulation for vaccine, *Pharm. Biotechnol.* **6**:313-324.

27. P.K. Hansrani, S.S. Davis, and M.J. Groves, 1983, The preparation and properties of sterile intravenous emulsions, *J. Parenteral Sci. & Technol.* **37**:145-150.

28. H. Endoh, Y. Hashimoto, Y. Kawashima, and Y. Suzuki, 1980, Agglutination microassay of hapten- or protein-modified liposomes using a multiple cell culture harvester, *J. Immunol. Methods* **36**:185-195.

29. H. Endoh, Y. Suzuki, and Y. Hashimoto, 1981, Antibody coating of liposomes with 1-ethyl-3-(3-dimethyl-aminopropyl) carbodiimide and the effect on target specificity, **44**:79-85.

30. J.S. Ellingson, and W.E.M. Lands, 1968, Phospholipid reactivation of plasmalogen metabolism, *Lipids* **3**(2):111-120.

31. M.E. Klegerman, P.L. Zeunert, Y. Lou, P.O. Devadoss, and M.J. Groves, 1993, Inhibition of murine sarcoma cell adherence to polystyrene substrata by Bacillus Calmette Guérin: Evidence for fibronectin-mediatiated direct antitumor activity of BCG, *Cancer Invest.* **11**(6):660-666.

32. F. Grinnel, D.G. Hays, and D. Minter, 1977, Cell adhesion and spreading factor: Partial purification and properties, *Exp. Cell. Res.* **110**:175-190.

33. K.M. Yamada, and D.W. Kennedy, 1984, Dualistic nature of adhesive protein function: Fibronectin and its biologically active peptide fragments can autoinhibit fibronectin function, *J. Cell. Biol.* **99**:29-36.

PHYSICAL CHARACTERIZATION AND STABILITY OF A MICROEMULSION FOR POTENTIAL ORAL ADMINISTRATION OF A PEPTIDE

Ali Türkyılmaz[1], Nevin Çelebi[1], Bilge Gönül[2], and Hayat Alkan-Önyüksel[3]

1 Department of Pharmaceutical Technology
  Faculty of Pharmacy
  Gazi University
  06330, Ankara, Turkey

2 Department of Physiology
  Gazi University Medical Faculty
  06510, Ankara, Turkey

3 Department of Pharmaceutics and Pharmacodynamics
  College of Pharmacy
  University of Illinois at Chicago
  Chicago, IL 60612

INTRODUCTION

Over the past few years, much attention has been paid to potential pharmaceutical uses of microemulsions as novel drug delivery systems. Microemulsions are defined as multicomponent systems consisting a unique ratio of component including a lipophilic phase, a hydrophilic phase, a surfactant and a co-surfactant[1]. Main characteristics of this system are low viscosity, isotropicity, thermodynamically stability[2] and droplet diameter less than 100 nm[3]. Such systems are formed spontaneously[3,4]. The use of microemulsions as possible therapeutic systems is interesting for two main reasons: a) controlled drug release[5], b) increased systemic and topical absorption of drugs[4,6,7].

A microemulsion formulation must provide a stable system which is less affected by different physiological states such as food, bile flow, pH etc[3]. A microemulsion formulation always has to be tailor-made according to the characteristics of the drug compound. Even slight changes in the chemical structure of the active molecule might affect the characteristics of the mixture up to a complete disappearance of the microemulsion structure[3].

Peptides generally are poorly absorbed from the gastrointestinal tract. Enhancing oral absorption of peptide drugs is very challenging due to the enzymatic barrier limiting the absorption of natural peptide drugs from the gastrointestinal tract[8]. Different strategies have been attempted to enhance absorption using enhancers[9], chemical modification[10] or formulation design[11].

Studies have been reported on the use of liposomes, lipid-surfactant micelles[12], oil vehicles[13], emulsions[14] and microemulsions[4,15] for drug delivery. It seems that microemulsions may be ideal drug delivery systems for steroids, diuretics, antibiotics, vitamins, antineoplastic agents, peptides (insulin, vasopressin, cyclosporine) etc.[1]. Especially, microemulsions have been recently suggested as carriers for peroral peptide drugs.

The objective of this study was to prepare a microemulsion formulation of epidermal growth factor (EGF), which prevents the enzymatic degradation in gastrointestinal systems. EGF is mitogenic polypeptide composed of 53 amino acids. In this part of the study, we report on our in vitro preliminary studies of the physical characteristics and stability of microemulsions.

MATERIALS AND METHODS

Labrafil® M 1944 CS (unsaturated polyglycolysed glycerides) was kindly provided from Gattefossé (France). Arlacel 186 (glycerolmonooleate-propylene glycol) and Brij 35 (polyoxyethylene lauryl ether) were supplied by ICI Pharmaceutics (England). EGF was purchased from Sigma (USA).

Preparation of the Microemulsion

Arlacel 186 and Brij 35 were used as surfactants (S), Labrafil® M 1944 CS was the oil phase. Absolute alcohol and distilled water were used as the co-surfactant (Co-S) and the aqueous phase respectively. Pseudoternary phase diagrams were established by titration. First, the ratio surfactant/co-surfactant was optimized using the following S/Co-S ratios: 1.0, 1.5, 2.0, 2.5 and 3.0. The area light by the microemulsion's existence field was calculated and plotted as a function of S/Co-S as shown in Figure 1. The microemulsions existence field for S/Co-S of 2.5 is shown in Figure 2. For the studies a microemulsion composition was chosen being in the approximate center of gravity of the microemulsion existence field.

Briefly, the aqueous and non-aqueous phases were prepared separately and then mixed with a stirring bar until a clear formulation was obtained. EGF was dissolved in the water phase and then added to the other phase containing S, Co-S and oil.

Physical Characterization of Microemulsions

The physical characteristics of the microemulsion were measured at different temperatures following (+4, 30 and 40 °C) during 12 months, unless otherwise indicated:

Phase Separation. The microemulsions were centrifuged at 1200 (g) for 5 hours to observe if phase separation occurred. The test was carried out only at room temperature.

Determination of the Type of Microemulsion. To determine if the microemulsion is of o/w or w/o type, the dye test (Sudan III- Merck) was performed.

Conductivity Measurements. Conductivity measurements were made using YSI Model 35 conductance meter. In each case, the measurements were made at 25 °C using a glass dipping cell with platinum electrodes. The cell constant was determined using standard KCl solution.

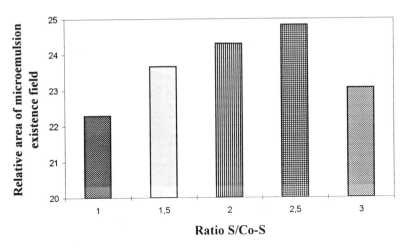

Figure 1. Relative areas of microemulsion existence field as function of S/Co-S ratio.

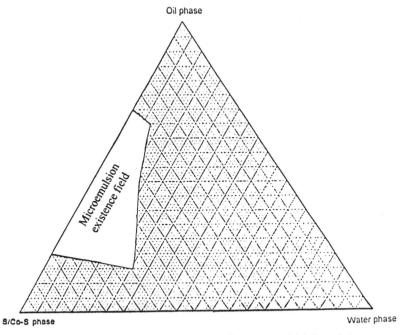

Figure 2. Pseudo-ternary phase diagramme for microemulsion formulation.

Refractive Index Measurements. Refractive index (RI) measurements were carried out using Shimadzu refractometer.

Droplets Size Measurements. The mean size of the emulsions droplets was determined by quasi-elastic light scattering (QELS) measurements, using a Nicomp Model 270 submicron particle sizer equipped with a 5 mW helium-neon laser at an exciting wavelength of 632.8 nm, 64 channel autocorrelation function, a temperature controlled scattering cell holder, and an ADM 11 video display terminal computer for analyzing the fluctuations in scattered light intensity generated by the diffusion of particles in solution. The hydrodynamic particle diameter, d_h, was obtained from the Stokes-Einstein relation using the measured diffusion coefficient obtained from the fit.

Viscosity Measurements. The viscosity of the microemulsion was measured using a capillary viscometer at 25 ± 0.1 °C (Ubbelohde viscometer). The instrument was calibrated with liquid of known viscosity. The viscometers were placed in a water bath at the constant temperature of 25.0 ± 0.1 °C.

Turbidity Measurements. Turbidity measurements were made using a Hach Model 2100 A turbidimeter. As a blank a test solution (0-100 NTU Range 16) was used in these measurements.

The Density and the pH Measurements. The density and the pH of microemulsion were also measured using picnometre and pH-Meter CG840 respectively.

All physical characteristics of the microemulsion were measured in the absence of epidermal growth factor. Statistical analysis were evaluated by one-way analysis of variance (ANOVA) test.

RESULTS AND DISCUSSION

As mentioned in the introduction, microemulsions have four components; water, oil and a surfactant in combination with a co-surfactant. The surfactant to be chosen should be suitable to prevent aggregation. In order to select the appropriate surfactants with low and high hydrophilic-lipophilic balance (HLB) values, Arlacel 186 (HLB:2,8) and Brij 35 (HLB:16.9) have been used in this study. The co-surfactant interacts at the interface to form a mixed duplex film. The co-surfactant also appears to act in a dynamic state.

Short chain alcohols, such as isopropanol, n-butanol and n-pentanol, were used as co-surfactant in microemulsion formulations[16]. Absolute ethanol was used as the co-surfactant in our study because of low toxicity for oral administration.

Density, pH and RI measurements are important and useful in identification testing of microemulsion formulations. Turbidity, conductivity and droplet size measurements provide powerful tools to investigate microemulsions and hold more promise as stability indicators. All these techniques were applied in the present study. Moreover, viscosity measurements were also carried out in an attempt to obtain some information on the thickness of the surfactant film.

The type of microemulsion was found to be W/O as determined by the dye method. The density of microemulsion was between 0.931 and 0.960 g/mL (Table 1) and the pH varied from 4.47 to 4.69 (Table 2). Both the density and the pH didn't change during the study period at different temperatures (p>0.05). According to centrifuge test no precipitate was observed and microemulsion remained completely transparent during storage temperatures.

Table 1. Density of microemulsions

Time (months)	4 °C Mean ± C.I. (g/mL)	30 °C Mean ± C.I. (g/mL)	40 °C Mean ± C.I. (g/mL)
0	0.950±0.010	0.950±0.010	0.950±0.010
1	0.946±0.011	0.954±0.026	0.942±0.030
2	0.960±0.003	0.958±0.019	0.950±0.012
3	0.945±0.012	0.958±0.019	0.939±0.013
4	0.943±0.016	0.940±0.012	0.931±0.025
5	0.942±0.017	0.940±0.014	0.934±0.010
6	0.949±0.008	0.944±0.014	0.934±0.015
9	0.937±0.026	0.940±0.012	0.933±0.038
12	0.945±0.013	0.940±0.014	0.935±0.012

C.I:Confidence interval ($t_{0.05}$, n=3)

Table 2. pH of microemulsions

Time (months)	4 °C Mean ± C.I.	30 °C Mean ± C.I.	40 °C Mean ± C.I.
0	4.47±0.28	4.47±0.28	4.47±0.28
1	4.61±0.04	4.59±0.10	4.67±0.19
2	4.60±0.14	4.57±0.14	4.58±0.04
3	4.69±0.25	4.69±0.21	4.69±0.25
4	4.63±0.06	4.65±0.12	4.62±0.04
5	4.62±0.05	4.64±0.10	4.62±0.05
6	4.62±0.06	4.67±0.07	4.65±0.08
9	4.58±0.06	4.63±0.05	4.68±0.05
12	4.59±0.10	4.63±0.08	4.67±0.11

C.I:Confidence interval ($t_{0.05}$, n=3)

Table 3. Turbidity values of microemulsions

Time (months)	4 °C Mean ± C.I. (NTU)	30 °C Mean ± C.I. (NTU)	40°C Mean ± C.I. (NTU)
0	11.3 ± 0.7	11.3 ± 0.7	11.3 ± 0.7
1	11.7 ± 0.4	11.3 ± 0.5	11.1 ± 0.4
2	11.8 ± 0.3	11.1 ± 0.4	11.1 ± 0.4
3	11.9 ± 0.5	11.1 ± 0.4	11.1 ± 0.4
4	11.6 ± 0.4	11.6 ± 0.4	11.1 ± 0.5
5	11.8 ± 0.3	11.6 ± 0.4	11.1 ± 0.5
6	11.9 ± 0.4	11.5 ± 0.3	11.1 ± 0.5
9	11.9 ± 0.4	11.6 ± 0.4	11.3 ± 0.7
12	11.8 ± 0.7	11.7 ± 0.4	11.4 ± 0.4

C.I:Confidence interval ($t_{0.05}$, n=3)

Table 4. Conductance of microemulsions

Time (months)	4 °C Mean ± C.I. (μmhos)	30 °C Mean ± C.I. (μmhos)	40 °C Mean ± C.I. (μmhos)
0	0.55 ± 0.07	0.55 ± 0.07	0.55 ± 0.07
1	0.51 ± 0.02	0.54 ± 0.01	0.55 ± 0.02
2	0.52 ± 0.05	0.50 ± 0.03	0.51 ± 0.05
3	0.50 ± 0.02	0.50 ± 0.02	0.54 ± 0.02
4	0.51 ± 0.02	0.50 ± 0.02	0.50 ± 0.02
5	0.51 ± 0.03	0.50 ± 0.02	0.51 ± 0.04
6	0.52 ± 0.03	0.50 ± 0.02	0.51 ± 0.03
9	0.51 ± 0.08	0.51 ± 0.04	0.51 ± 0.04
12	0.50 ± 0.06	0.52 ± 0.07	0.50 ± 0.05

C.I:Confidence interval ($t_{0.05}$, n=3)

Table 5. Refractive index of microemulsions

Time (months)	4 °C Mean ± C.I.	30 °C Mean ± C.I.	40°C Mean ± C.I.
0	1.4420 ± 0.0050	1.4420 ± 0.0050	1.4420 ± 0.0050
1	1.4419 ± 0.0000	1.4429 ± 0.0001	1.4415 ± 0.0001
2	1.4429 ± 0.0000	1.4440 ± 0.0020	1.4425 ± 0.0001
3	1.4430 ± 0.0000	1.4430 ± 0.0000	1.4423 ± 0.0015
4	1.4418 ± 0.0050	1.4400 ± 0.0000	1.4413 ± 0.0006
5	1.4435 ± 0.0001	1.4435 ± 0.0001	1.4433 ± 0.0000
6	1.4435 ± 0.0001	1.4427 ± 0.0060	1.4434 ± 0.0001
9	1.4433 ± 0.0001	1.4500 ± 0.0250	1.4434 ± 0.0001
12	1.4435 ± 0.0001	1.4442 ± 0.0040	1.4435 ± 0.0001

C.I.:Confidence interval ($t_{0.05}$, n=3)

Table 6. Apparent viscosity of microemulsions

Time (months)	4 °C Mean ± C.I. (mPas)	30 °C Mean ± C.I. (mPas)	40°C Mean ± C.I. (mPas)
0	41.3 ± 1.0	41.3 ± 1.0	41.3 ± 1.0
1	41.1 ± 0.6	40.1 ± 0.7	39.7 ± 0.5
2	46.1 ± 0.2	44.8 ± 0.3	44.3 ± 0.3
3	47.5 ± 0.2	47.3 ± 0.1	46.1 ± 0.2
4	47.4 ± 0.6	47.1 ± 0.2	47.1 ± 0.1
5	47.5 ± 0.4	47.1 ± 0.4	47.0 ± 0.4
6	47.8 ± 0.1	47.2 ± 0.1	46.9 ± 0.3
9	47.8 ± 0.3	48.2 ± 0.1	46.8 ± 0.4
12	48.4 ± 0.7	47.7 ± 0.5	46.9 ± 0.2

C.I.:Confidence interval ($t_{0.05}$, n=3)

The mean size of the microemulsion droplets was approximately 6-10 nm and remained constant through out the experiment. The results of turbidity are shown in Table 3. In general, the droplet size of microemulsions increased as turbidity increased. The results indicated that, the microemulsion investigated have been stored at different temperatures for up to 12 months and no significant changes in both droplets size and turbidity ($p>0.05$).

Table 4 summarizes the conductivity measurements. A significant difference wasn't found for the conductivity of the microemulsion at different temperatures during 12 months. It is possible to relate the difference in conductivity behavior to the possible structure of the system. Conductivity results point toward the presence of definite water cores with a surfactant layer which act as a barrier for ion transport[17].

Table 5 summarizes the RIs measurements. A significant difference wasn't found in RI of microemulsions at different temperatures during 12 months period ($p>0.05$).

Table 6 shows the viscosity of the investigated W/O microemulsions at different temperatures. It was observed that, the viscosity of the microemulsion changed slightly with time at storage temperatures ($p<0.01$). This change is due to evaporation of the alcohol present in the outer phase of the microemulsion during measurements.

CONCLUSION

It was concluded that the physical characteristics of the developed microemulsion, didn't change under different storage temperatures. Thus we have succeeded in preparing a stable microemulsion for potential oral applications of EGF. In addition, chemical stability of microemulsion containing EGF was investigated and observed that EGF was stable in the microemulsion.

Acknowledgments

This work was supported by Turkish Scientific and Technical Research Council (TÜBİTAK). The authors wish to thank Gattefossé (France) for the generous supplies of chemicals.

REFERENCES

1. H.N.Bhargava, A. Narurkar, and L.M. Lieb, Using microemulsions for drug delivery, *Pharm. Technol.* 11(3): 46-54(1987).
2. J.M.Sarciaux, A.R. Hilgers, A.M. Cooper, K.J. Cook, P.S. Burton and M.V. Patel, Oral absorption potential of peptidic drugs from microemulsion formulations: A case study of U-71038, *B.T. Gattefossé, 88: 11-19*(1995).
3. A.Meinzer, E. Mueller and J. Vonderscher, Microemulsion a suitable galenical approach for the absorption enhancement of low soluble compounds, *B.T. Gattefossé, 88: 21-26*(1995).
4. W.A.Ritschel, Microemulsions for improved peptide absorption from the gastrointestinal tract, *Meth. Find. Exp. Clin.Pharmacol.*13 (3):205-220(1991).
5. R.Wallin, H.Dyhre, S.Björkman, A.Fyge, S.Engström and H.Renck, Prolongation of lidocaine induced regional anesthesia by a slow-release microemulsion formulation, *Proceed.Int'l. Symp. Control. Rel. Bioact. Mater.*, 24, 555-556(1997).
6. P.P.Constantinidis, J-P. Scalart, C. Lancaster, J.Marcello, G. Marks, H. Ellens and P.L. Smith, Formulation and intestinal absorption enhancement evaluation of water-in-oil microemulsions in corporating medium-chain glycerides, *Pharm. Res.*,11(10),1385-1390 (1994).
7. F.Février, M.F. Bobin, C. Lafforgue and M.C. Martini, Advances in microemulsions and transepidermal penetration of tyrosine, *S.T.P. Pharma Sciences*, 1 (1):60-63(1991).
8. V.H.L. Lee, S. Dodda-Kashi, G.M. Grass and W. Rubes, Oral route of peptide and protein drug delivery, in:Peptide and Protein Drug Delivery, V.H.L. Lee, ed., Marcel Decker Inc., New York, 691-738(1991).
9. V.H.L. Lee, Changing needs in drug delivery in the era of peptide and protein drugs, in:Peptide and Protein Drug Delivery, V.H.L. Lee, ed., Marcel Decker Inc., New York, 1-56(1991).
10. V.H.L. Lee, R.D. Traver and M.E. Taub, Enzymatic barriers to peptide and protein drug delivery, in: Peptide and Protein Drug Delivery, V.H.L. Lee, ed., Marcel Decker Inc., New York, 303-358(1991).
11. M.Lynda, M. Sanders, Controlled delivery systems for peptides, in: Peptide and Protein Drug Delivery, V.H.L. Lee, ed., Marcel Decker Inc., New York, 785-806(1991).
12. K.Takada, N.Shibata, H.Yoshimura,Y.Masuda, H.Yoshikawa, S.Muranishi, and T. Oka, Promotion of the selective lymphatic delivery of cyclosporin A by lipid-surfactant mixed micelles, *J.Pharmacobio-Dyn.* 8: 320-323(1985).
13. J.P.Reymond, H. Sucker and J. Vonderscher, In vivo model for cyclosporin intestinal absorption in lipid vehicles, *Pharm.Res.*, 5: 677-679(1988).
14. B.D.Tarr, and S.H.Yalkowsky, Enhanced intestinal absorption of cyclosporine in rats through the reduction of emulsion droplet size, *Pharm.Res.*, 6: 40-43(1989).
15. W.A.Ritschel, S. Adolph, G.B. Ritschel and T. Schroeder, Improvement of peroral absorption of cyclosporine A by microemulsions, *Meth.Find.Exp.Clin. Pharmacol.*, 12(2), 127-134(1990).

16. R.Aboofazeli and M.J. Lawrence, Incestigations into the formation and characterization of phospholipid microemulsions. I.Pseudo-ternary phase diagrams of systems containing water-lecithin-alcohol-isopropyl myristate, *Int.J.Pharm.*, 93,161-175(1993).

17. R.C.Baker, A.T.Florence, R.H. Ottewill and F.Tadros, Investigations into the formation and characterization of microemulsions II. Light scattering conductivity and viscosity studies of microemulsions, *J.Coll.Int.Sci.*, 100(2), 332-349(1984).

UREA PERMEATION THROUGH COMPLEX COACERVATE MEMBRANES

Sümer Peker, Şerife Helvacı, and Handan Esen

Ege University
Chemical Engineering Department
35100 Bornova
İzmir, Turkey

INTRODUCTION

The facility with which urea forms hydrogen bonds makes it a frequently used component in many cosmetic applications as a moisturizer and exfoliator. Encapsulation of sponges containing urea in gelatin/gum arabic complex coacervates could provide additional resistance for the regulation of its release rate[1]. Besides its practical significance, diffusion of urea through gelatin/gum arabic complex coacervates is also of interest due to its interaction with the polymer network of the coacervate, exemplifing a case of mass transfer with chemical reaction. Urea interacts with the coacervate network by breaking the hydrogen bonds existing between gelatin and/or gum arabic segments[3] and forming these bonds with its own O and H atoms. The process is illustrated schematically in Figure 1 (a) and (b).

Urea also reacts with formaldehyde available for reaction within the membrane: In a previous study[2] it was found that only a fraction of the formaldehyde added to the coacervating system was actually used in the cross-linking reactions due to steric hindrance during the coacervation process. The results of the work could be formulated with a regression coefficient of $r^2 = 0.9978$ by the equation,

$$y = 0.8312 \, x^{3.98} \tag{1}$$

where,
$y = $ log (formaldehyde/gelatin)$_{actual}$
$x = $ log (formaldehyde/gelatin)$_{input}$
signifying the weight of formaldehyde per unit weight of gelatin added to the system (x) and that which is actually used (y) in the crosslinking reactions. Formaldehyde not consumed in the cross - linking reactions either remain in free state within the interstices of the network or as unbound monomethylol end groups which could not form cross-links. In the first case it forms resins with the diffusing urea and in the second case it forms cross links. The cross-linking reaction is shown schematically in Figure 1(c) and (d). Resin formation reactions can be represented with the following formula in Figure 1.

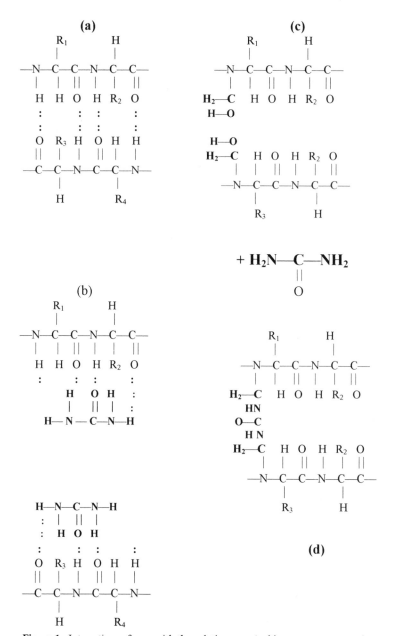

Figure 1. Interactions of urea with the gelatin - gum arabic coacervate network.

$$n(NH_2CONHCH_2OH) \rightarrow H(NHCONHCH_2)_n\ OH + (n-1)H_2O \qquad (2)$$

These two interactions have opposing effects on the coacervate network: In the case of chemical reaction, both the cross-linking and the resin formation result in a more compact polymer network, increasing the resistance to mass transfer. In the case of breakage of hydrogen bonds, the polymers tend to become more linear under the action of tensile stresses in dry state and osmotic pressures in a solution medium, elongating in the former and swelling extensively in the latter case. Swelling increases the free area for mass transport facilitating the diffusion of solutes. In this work, diffusion of urea is investigated as a function of concentration differences between the two sides of coacervate membranes and thickness of the membrane.

MATERIALS AND METHODS

Materials

Gum arabic, gelatin and urea used in this work were purchased from Merck. The other reagents, obtained from local firms, were more than 99% pure. Formaldehyde was obtained as 37% solution.

Preparation of the Membranes

Coacervation is a well known process where equal amounts by weight of gelatin and gum arabic are attracted by electrostatic charges to form a continuous gel medium. When the concentration of the polymers is reduced to less than 3% of the solution medium, the polymer gel settles down as a coacervate or colloid-rich phase. Formaldehyde may be added at this step to form permanent linkages between the polymers. The following procedure was adopted in this work: 2%(w) solutions of gum arabic and gelatin were mixed at 40 °C to obtain a homogeneous solution. pH was adjusted to 4 by addition of acetic acid. 37%(w) formaldehyde solution is added in an amount proportional to the degree of crosslinking desired in the coacervating gel. The coacervate settles to its equilibrium height after 24 hours. The colloid poor equilibrium phase is discarded after determination of its nitrogen and formaldehyde contents. The cast membrane is washed by soaking in water to remove the unreacted components and then left to dry at room temperature free of convection currents. Membranes of the same diameter but of two different thicknesses, 0.4 and 0.8 mm (dry) were prepared in this work. Since no plasticizers are used, the membranes become brittle when dried. They were kept in a 90 % r.h. chamber until they attained an equilibrium moisture of 12.5 % before being attached to the diffusion cell.

The Diffusion Cell

The diffusion cell, shown schematically in Figure 2, consisted of two compartments each of 2.5 L capacity, separated by a plate which had a centrally placed hole of 5 cm diameter. The membrane extended over this hole and was fixed by annular clamps and flanges. One of the cells (A) was filled with urea solutions of concentrations 0.230, 0.326, 0.604 or 0.730 mol/L and the other with water or 2%(w) formaldehyde solution. Formaldehyde solutions were used to confirm the effect of entrained formaldehyde in the case of urea diffusion to the water compartment. Both compartments were stirred at a rate of 380 rpm to eliminate the formation of concentration gradients at both sides of the membrane. Samples were taken at intervals from both compartments for analysis of the urea and formaldehyde contents.

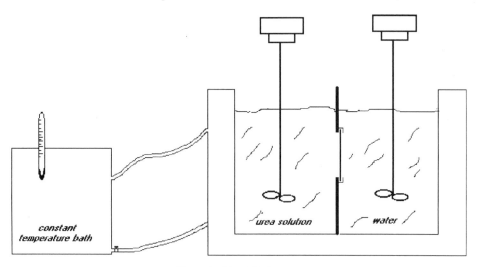

Figure 2. The diffusion cell.

Determination of the Urea and Formaldehyde Contents

Amount of urea was determined in terms of its total nitrogen content by a micro-kjeldahl method[4]. Formaldehyde content of the samples were determined by titration with NaOH in the presence of hydroxyl amine hydrochloride with bromophenol blue as the indicator[5].

Mathematical Analysis of the Results

Initial and final weights, thicknesses and formaldehyde contents of the membranes and the variation with timeof urea and formaldehyde concentrations in both compartments were known in the experiments. From these known values the diffusion coefficients, concentration and mass flux profiles within the membranes were found by numerically solving the unsteady state mass transfer equation, with terms added to account for the chemical reaction and moving boundaries. The results of this analysis are discussed elsewhere[6].

RESULTS

Experimental conditions and the overall results obtained are given in Table 1. Initial values of the concentrations in the two compartments (cells) on either side of the membrane and its initial thickness in the dry state are given in the first three columns, respectively. Duration given in the 4th column signifies the time it took for the membrane to rupture. % swelling values given in the last column show the ratio of the final thickness to the initial (dry) thickness of the membrane.

Release rates are given as % released versus time in Figure 3. "% released" term denotes the ratio of the total quantity transferred, to the initial quantity present in the A (urea solution) compartment. It is a gross quantity including both the amount retained within the membrane and the amount actually transferred to the B (water or formaldehyde solution) compartment, during the first 60 minutes, if it did not rupture before that time. 60 minute reference time was chosen to compare the different conditions on the same basis.

Total release rates up to the time the membrane ruptured are given in Figure 4, as a function of the initial concentration in the urea solution compartment, for different film thicknesses and for the case of formaldehyde counter-diffusion.

Table 1. Experimental conditions and results

Concentration in cells (mol/L)		Thickness of Film (mm)	Duration (min)	% swelling
A (urea)	B(water)			
0.230	0	0.8	80	92
0.326	0	0.8	116	100
0.604	0	0.8	140	101
0.730	0	0.8	60	40
0.326	0	0.4	60	23
0.604	0	0.4	40	44
Experiments performed to confirm the effect of formaldehyde				
0.230	2%	0.4	30	50
0.326	2%	0.4	30	25
0.604	2%	0.4	30	31
0.730	2%	0.4	30	7

When the pH of the medium is greater than five (neutral or slightly basic medium) formaldehyde diffuses freely into the urea compartment without entering into any kind of reaction. In the case of water in the B compartment, the concentration driving force for formaldehyde transfer is equal in both directions. So a concentration difference indicates the direction of swelling and/or barrier formation in the membrane. The results are summarized in Table 2 in terms of weight % formaldehyde.

The results clearly show that when there is only water in the B compartment, a barrier forms in the direction of the urea compartment within the membrane. When there is a formaldehyde solution in the B compartment, concentration of formaldehyde in the A (urea) compartment increases with an increase in the initial urea concentration, indicating the extent of swelling due to hydrogen bond breakage in that direction.

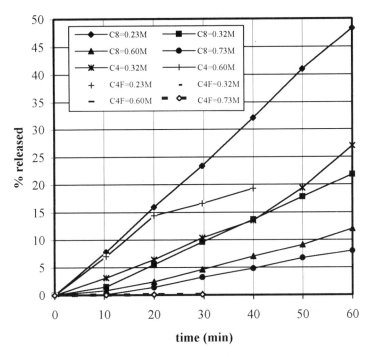

Figure 3. % Released as a function of time, for different membrane thicknesses concentration driving forces and formaldehyde concentrations.

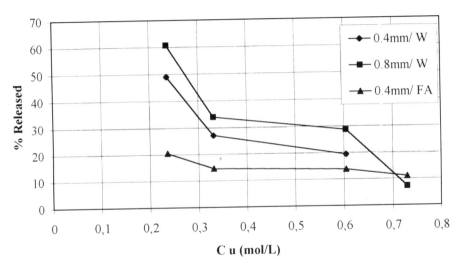

Figure 4. % released as a function of the initial urea concentration (concentration driving force).

DISCUSSIONS

The results presented above clearly point out that the net rate of mass transfer, e.g., the % release, is an interplay of the barrier effect of the urea/formaldehyde and monomethylolurea/formaldehyde reactions and the hydrogen bond breaking action of urea. The constant slopes of the linear profiles observed in Figure 3 show that rate of mass transfer, does not change (e.g., remains constant) in spite of high swelling ratios and decreased resistances within the membrane. In addition the slopes (rates) decrease as the initial urea concentration in the A compartment increases. These two phenomena can only be possible if a barrier forms within the membrane which acts as a bottle neck in determining the rate of mass transfer. This barrier shifts toward the formaldehyde compartment, when the concentration of formaldehyde in that direction is greater than the concentration of the entrained formaldehyde within the membrane.

The increase in the formaldehyde concentration in the urea compartment with an increase in the initial urea concentration suggests that urea functions somewhat like a plasticizer within a solution medium. The elongational behavior of the composite coacervate films under tensile stresses were investigated in a previous study[7]. The results for coacervates are summarized in Table 3.

Table 2. Formaldehyde concentrations in the compartments at the end of the experiments (weight % formaldehyde)

	8 mm membrane/ water				4 mm membrane/ water			4 mm membrane/ formaldehyde			
C_u (M)	0.23	0.32	0.60	0.73	0.23	0.32	0.60	0.23	0.32	0.60	0.73
A	0.37	0.41	0.46	0.60	0.32	0.39	0.51	0.37	0.40	0.43	0.49
B	0.40	0.50	0.85	0.92	0.41	0.46	0.55	1.07	1.19	1.27	1.33

Table 3. Effect of plasticizers on the tensile strength and elongational behavior of coacervate membranes

% plasticizer	C_{FA} (mol/L)	$E_{2\%}$	TS (MN/m^2)	ε_b %	$E_{cr}/E_{2\%}$
20	0	2.2	5.3	51.3	4.82
20	0.245	340	6.8	2	0.08
40	0.245	51	9.6	56	0.26

The first row gives the uncrosslinked, while the second row gives the crosslinked membrane behavior. Third row gives the effect of a relative increase in the concentration of the plasticizer. Young's modulus of elasticity at 2% elongation, $E_{2\%}$, indicates the rigidity of the membranes, while TS and $\varepsilon_b\%$ denote the tensile strength and elongation at break, respectively. The ratio $E_{cr}/E_{2\%}$ gives the crystallinity or the extent of the parallel conformation of the polymer chain segments. The table shows that an increase in the plasticizer concentration decreases the rigidity, and increases the strength, the rate of elongation and the the extent of the parallellism between the chain segments. To demonstrate the analogy between the two phenomena, the rupture times of the membranes are given as a function of the degree of swelling in Figure 5. Following the same arguments, the osmotic pressure developed within the membrane in aqueous media corresponds to the tensile stresses and elongation, to the degree of swelling. The short rupture times in the case of thin membranes exposed to high levels of urea concentration and in the case of counter diffusion of formaldehyde are due to the rigidity and the inhomogeneous stress distribution within the network.

The large life times of the thick membranes may also be due to intra molecular cross-link formation, instead of inter molecular bond formation as in the case of thin films. This is to be expected as twice the volume of solutions are subjected to the same amount of energy input (stirring rate, intensity and duration) during the formation of the coacervate. In case this energy is not sufficient to disperse the macromolecules, intra molecular links are formed in the presence of formaldehyde, leaving long molecular segments between highly cross-linked segments to stretch out in the presence of osmotic pressures.

CONCLUSIONS

The main results obtained in this work can be summarized as follows:
Net rate of urea release from the gelatin-gum arabic coacervate membranes is an interplay between the extent of chemical and physical interactions of urea with the polymers and the entrained formaldehyde.

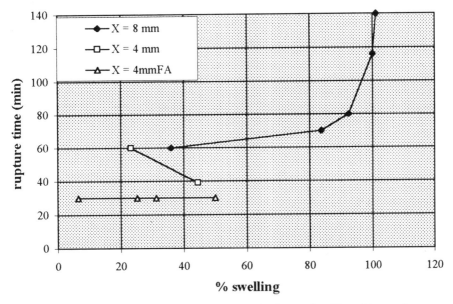

Figure 5. The relation between the rupture time and the degree of swelling of the membranes.

Initial segregation during coacervation within the membrane leads to local inhomogeneities which cause excessive swelling, increased release rates and early rupture of the films before equilibrium can be established.

Extent of swelling and the stability (rupture time) of the membrane have a direct bearing on the elongation and strength of the dry films under tensile stress.

REFERENCES

1. H. Yoshizawa, Y. Uemura, K. Ijichi, T.Hano, Y. Kawano, Y. Hatate, 1996, Permeability control of active agent from polymeric microcapsules by coating of gelatin/gum arabic membrane, *J. Chem. Eng. Japan* **29**:379-381.

2. S. Peker-Başara, B. Övez, İ. Balcıoğlu, 1993, Properties of gelatin-gum arabic coacervates composited with amino resins, *J. Chem. Tech. Biotechnol.* **56**:175- 184.

3. C.T. Greenwood, E.A.Milne, 1968, Natural High Polymers, Oliver & Boyd, Edinburgh

4. Horwitz, W., 1975, Methods of analysis of the association of analytical chemists, Analytical Chemists Publ., Washington, pp.327-328.

5. Kline, M.G., 1966, Analytical chemistry of polymers, Interscience Publishers, Inc., New York, pp. 64-66.

6. Ş. Helvaci, H. Esen, S. Peker, Diffusion of urea through gelatin - gum arabic membranes, Paper reviewed and in the process of revision for publication in: *J Membrane Science.*

7. N. Böke, B.Övez, S.Peker, 1994, Elastic behavior of resin /coacervate composites, *Engineering Systems Design and Analysis, ASME,* **64 (2):** 159-163

BACTERIAL POLYHYDROXYALKANOATES: BIOSYNTHESIS, SCREENING AND CHARACTERISATION

İhsan Gürsel and N.Gürdal Alaeddinoğlu

Middle East Technical University, Department of Biological Sciences, Biotechnology Research Unit, 06531, Ankara, TURKEY

INTRODUCTION

Plastic products are a part of our everyday life, but there are rising concerns worldwide that petroleum-based plastics are having detrimental effects on the environment. These products have estimated degradation times ranging from 20 years for vending machine cups, to 100 years for PET plastic bottles and 500 years for polystyrene foam caps and dishes. Although recycling is improving, about 75% will still be disposed into landfills and new strategies that can overcome this environmental problem will need to be investigated.

Advances in polymer chemistry, microbiology and molecular biology have made possible a variety of materials of biological origin to become available and used specifically in medicine and in other fields. These materials are collectively referred to as biodegradable polymers. Among many alternatives, two polymers showing great long-term promise are polylactides and polyhydroxyalkanoates (PHAs).

Polyhydroxyalkanoates (PHA) are a family of bacterial polyesters stored as intracellular granules by various aerobic and anaerobic microorganisms (Gürsel and Hasırcı, 1995b) in response to nutrient limitation (Table 1). It was first discovered by Lemoigne (Lemoigne, 1926), and since then the knowledge in this field has greatly expanded. Polyhydroxyalkanoic acids comprise almost 40 different 3-, 4-, and 5-hydroxyalkanoic acids (Anderson and Dawes, 1990; Steinbüchel and Valentin, 1995), of which poly (3-hydroxybutyric acid, PHB) is the most abundant and best known example (Howells, 1982). These polymers are synthesized by bacteria and are deposited as cytoplasmic granules (inclusion bodies) if a suitable carbon source is provided to the cells in excess and if growth is impaired because of the lack of one essential nutrient or oxygen (Muller and Seebach, 1993; Lee, 1996; Brandl et.al., 1990). These classes of biological macromolecules serve as an intracellular storage material to be used in nutrient depletion (Doi, 1990).

PHA can be recovered from bacterial cells following, large-scale fermentation by simple solvent extraction or enzymatic processes (Lee, 1996). PHA possesses physical and mechanical properties that are competitive with the existing high tonnage petroleum-based polymers such as polyethylene and polypropylene (Doi, 1990; Holmes, 1985). Polyhydroxybutyrate and its copolymers (with varying co-monomer types and compositions) has several unique physicochemical features such as thermoplasticity, biodegradability, biocompatibility, stereospecificity, optical activity, and non-toxicity (when used *in vivo)* (Gürsel, 1995). PHB's high thermoplasticity and low degradation rate, in the presence of hydrolytic enzymes, are presently the unfavourable properties making the product unsuitable for various applications (such as in pharmacy, medicine and in cosmetics as a drug delivery device construction material). However, when the copolymers containing hydroxyvalerate (HV) with varying molar concentrations are used, the crystallinity of material decreases approximately by 50% modifying the bulk properties of the material, including its degradability (Muller and Seebach, 1993, M. Yasin, 1987, 1992). When PHA is exposed to microorganisms of the ecosystem, degradation to carbon dioxide and water is achieved in about 1 to 6 months depending on the PHA variant. The carbon dioxide does not contribute to greenhouse problems as fixed by plants to glucose, and can thus be used for further PHA production. The cycle is truly sustainable. Furthermore the bacterial product similarly is degradable when used in vivo (controlled release devices etc.).

PHB and PHBV is presently produced commercially (300 tons/pa) by Zeneca Bioproducts under the trade name "BIOPOL™" (recently becoming a Monsanto subsidiary). The cost is approximately 20$ per kilo (1995 price), and is uncompetitive with existing petroleum products. Further exploitation of this technology for commodity uses will require a significant reduction in cost. In order to optimise the bacterial growth conditions leading to high production yields of tailor made polymers of desirable physicochemical properties biotechnological approaches are unavoidable.

Table 1. The accumulation of PHA in a variety of microorganisms (Gürsel, 1995)

Genus	Carbon source	PHA content[a]	Stress condition
Acinetobacter	Acetate	12	Sulfate
Alcaligenes	Fructose	96	Nitrogen
Azospirillum	Malate	75	Phosphate
Azotobacter	Glucose	70	Oxygen
Bacillus	Propionic acid	53	Nitrogen
Beijerinkia	Glucose	74	Nitrogen
Clostridium	TPG broth	13	ND[b]
Escherichia	Glucose	95	Nitrogen
Halobacterium	Glucose, Starch	60	Phosphate
Metylobacterium	Methanol	60	NS[c]
Methylocystis	Methane	70	NS
Micrococcus	Glucose	21	NS
Nocardia	Butane	14	NS
Pseudomonas	Gluconate	78	Nitrogen
Rhizobium	Mannitol	70	ND
Rhodobacter	CO, and H_2	28	Oxygen
Rhodococcus	Pentenoate	53	Nitrogen
Rhodospirillum	Acetate	67	Nitrogen
Syntrophomonas	Acetate	5.8	Nitrogen

[a] denotes (%) dry weight

[b] not determined

[c] not specified

Alcalegenes latus and *Alcalegenes eutrophus* are commonly used bacteria; being strictly respiratory energy metabolism, they utilise CO_2, fructose, gluconate and various other carbon sources (ie.organic acids). Carbohydrates are degraded via Entner-Doudoroff pathway to pyruvate, which is oxidatively decarboxylated to Acetyl coenzyme A (acetyl-CoA). Under suitable conditions (ie. aerobic conditions) with well-adjusted nutrients, the cells synthesizes acetyl-CoA completely to CO_2 via TCA cycle. With limited oxygen supply and one of the essential nutrients depleted, the cells synthesize P(3HB) via a three-step PHA biosynthetic pathway. β-ketothiolase, an NADH-dependent acetyl-CoA reductase, and a PHA synthetase catalyzes the conversion of acetyl-CoA to acetoacetyl-CoA, and then to D(-)-3-hydroxybutryl-CoA, and subsequently to P(3-HB) (Brandl et.al., 1990; Pouton and Akhtar, 1996, Muller and Seebach, 1993; Slater et.al., 1988)

Several bacterial species are able to incorporate monomers of carbon chain lengths (C-3 to C-14) into their PHA as a pendant chain. This difference is likely to be due to the PHA synthetase of these microorganisms (Steinbuchel and Wiese 1992; Timm and Steinbuchel, 1990). The carbon source cost is an economic factor that is pivotal to industrial PHA production (Collins, 1987, Yamane, 1993). Therefore, the PHA yield from the carbon source is a factor that must be considered when the strains are screened for industrial processing.

To test whether the isolated organisms are PHA-producers, specific staining techniques (Sudan black or a basic oxazine dye "Nile blue A") are employed. The stained cells are then investigated under ordinary or fluorescence microscope. The inclusions appear under ordinary light microscope as black granules or as orange fluorescent spots under the fluorescent microscope.

Present discussion focuses to PHA rapid screening; a novel method developed that can be used to detect PHA content from the cells on-line by employing a fluorescent dye. The production of versatile forms of PHA has been also achieved by modifying the ingredients of the culture media. *A.latus* and *A.eutrophus* strains have been used in experiments involving single and double-step polyester biosynthesis respectively.

CULTURING CONDITIONS OF PRODUCER STRAINS

Alcaligenes latus and *Alcaligenes eutrophus* (ATTC 29713 and 17699) strains were employed in PHA production. The culture media were prepared as described by Doi (1990) and Shubert et. al., (1988).

Polyester production using *A.latus* was carried out by one-stage cultivation in which cell growth and polymer production occur simultaneously (Scheme 1a). The growth and production period was maintained at 42 h., and co-substrate addition was adjusted 19 h post inoculation, corresponding to the beginning of exponential phase (data not shown).

Experiments with *A.eutrophus,* involving two-stage cultivation (Scheme 1b), were carried out with cells that were first grown in nutrient-rich medium (for 25 h, at 30°C with aeration) and transferred aseptically into polymer production medium (incubated for further 48 h where nitrogen limitation was used as the stress condition to stimulate polyester synthesis). Valerolactone (VL), bromopropionic acid (BrPPA), and 4-hydroxybutyric acid (4-HBA) sodium salt were added to the cultures to promote different polyester biosynthesis.

ISOLATION OF POLYHYDROXYALKANOATES

The cells were harvested by centrifugation, washed with water and freeze-dried. To extract accumulated polyesters, the cells were refluxed for 12 hours in hot chloroform (15 mL/g dry cell). The solution was filtered to remove the cell debris and the chloroform solution was first concentrated in rotary evaporator and then introduced dropwise into cold ethanol (chloroform:ethanol; 1:10). The milky coloured liquid was mixed overnight on a magnetic stirrer at ambient temperature. The white coloured precipitate was dried under vacuum at 50°C and weighed to record the isolated PHA amount.

Scheme 1. Widely accepted routes of achieving polymer formation in microrganisms:

c) Parallel process (one stage) :

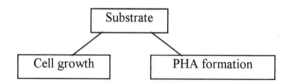

The substrate is utilised simultaneously for both cell growth and PHA formation

d) Serial process (two stage) :

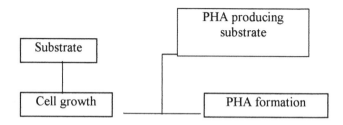

Polymer producing substrate is added to the culture at the onset of stationary phase

DETERMINATION OF PHA

Spectrophotometric detection or Nile blue A (a fluorescence dye) staining methods have been employed for polymer production and quantification (Senior et.al., 1972; Ward and Dawes, 1973; Williamson and Wilkinson, 1958; Law and Slepecky, 1961). The former involves spectrophotometric assay of the amount of crotonic acid produced from the cells, where absorbance values are obtained at 235nm and calculated from the extinction coefficient (1.55×10^4). The latter involves the use of Nile blue A. PHB producing cells, when stained, appear to have orange colored inclusions under fluorescent microscope. Apart from the usual extraction methods, here in, a new screening method is described, allowing reliable, quick and easy detection of the polmeric inclusions within the cells.

Crotonic Acid Assay

3-hydroxybutyric acid (0.5 mg/ml) monomer was dissolved in ethanol and diluted to 50μg/mL in ethanol. Two fold dilutions of standards (5-50μg/ml) were prepared. 0.1g and 0.2 g of dry cells of *A.eutrophus* were used in the determination PHA content. A calibration curve was constructed by using 6 standard solutions (5-50μg/ml). The absorbances of the standards were read at 235nm. The O.D results of the 0.1g and 0.2g dry cell samples were obtained.

Based on this assay, PHA content/g.dry cell weight for 0.1g.sample was 352.2mg, and for 0.2g sample was 311.1/mg, (equivalent to 53.2 % and 31.1 % recovery respectively) the average of these values is 33.15 % recovery form the producer strain.

To check the sensitivity of the assay, the remaining portion of the dry cells were studied by the usual extraction process (refluxing in CHCl₃ and purifying the polymer by precipitating it in ethanol). The results indicated that the cells deposited 36.4% polymer, coresponding to ca. 90 % sensitivity of the assay with respect to usual extraction process.

DEVELOPMENT OF A NOVEL METHOD FOR DIRECT ASSESMENT OF PHA PRODUCTION

Nile red is a fluorescent dye specific to intracellular lipids, thus allowing the PHB granules to be stained within the cells. The staining intensity is expected to show a linear relationship with increasing inclusion content. The dye can easily penetrate the cells and fluorescence intensity of the stained cells can be measured within a short period (with respect to standard procedure). Non-PHB producer strain staining is expected to give none or very low fluorescent measurement since the bacteria have almost no lipid inclusion granules. To demonstrate this, a non-PHB producing negative control strain, *E.coli* K12 was used and the fluorescence intensities were measured for both producer and non-producer strains (Figures 1 and 2, Tables 2 and 3).

Prior to staining of the bacteria, excitation and emission wavelenghts of the dye in solution was determined and a calibration curve was constructed by a spectrofluorimeter. Emission λ maximum (λem) and excitation λ maximum (λex) of the dye were found to be 616.0 nm and 572.8 nm respectively.

Staining of Cells with Nile Red and On-Line PHA Detection

λem and λex were determined for the stained cells as 569.6nm and 515nm. *Alcaligenes eutrophus* cells were harvested from 50ml growth medium and *E.coli* cells were from 20 ml nutrient broth (Difco) supplemented with yeast extract. The cells were fixed in ethanol, the pellets were then dissolved in 5 ml dH₂O. 1 ml from this solution was mixed with 1 ml of stock Nile red solution (0.2mg/ml ethanol) and the volume was completed to 5 ml by the addition of 3 ml of 70% ethanol. The solutions were incubated at room temperature for 2 hours followed by centrifugation. The pellets were used to determine the extent of staining, following suspension in 3 ml dH₂O. From these, 1 ml was taken and diluted as required (1/50, 1/100, 1/200, 1/250) and then calibration curves were plotted for *A.eutrophus* and *E.coli*. As seen from the calibration curves the fluorescence intensities were quite different for *E.coli* and *A.eutrophus* cells (compare Figures 2 and 3). Comparatively small amount of staining in *E.coli* can be attributed to the non-specific lipids and trace amounts of polymer found in the membrane of *E.coli* (Reusch and Sadoff, 1983, 1988).

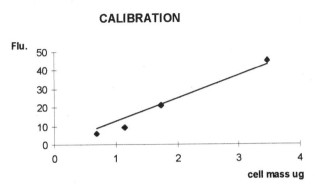

CALIBRATION

Figure 1. Calibration curve for the stained non PHB producing *E.coli* cells.

Table 2. Fluorescence measurement of Nile red
stained *E.coli* cells

Cell mass (mg)	Fluorescence Intensity
0.69	5.841
1.15	9.402
1.74	21.019
3.47	45.245

Table 3. Fluorescence measurement of Nile red
stained *A.eutrophus* cells

Cell mass (mg).	Fluorescence Intensity
0.78	123.375
0.99	157.046
1.31	204.619
1.97	295.379
3.94	626.029

Calibration curve of the Nile red stained cells was constructed by plotting fluorescence intensity vs cell mass. The unknown samples could be obtained by using the slope of this plot. To test this, a 10 ml aliquot (containing 50 mg dry cells) of previously isolated cells (grown on Fructose + 4HBA) were fixed in ethanol, stained and the fluorescence was measured. 0.048 and 0.193 cell concentrations gave 115.53 and 479.72 fluorescence intensities, corresponding to 119.7 and 497 1 µg PHA respectively. The PHA content was 21.4 mg for 50 mg dry cell. This was equivalent to polymer content of 42.8 % PHA. The usual extraction for this type of polymer gave 42.6 % yield.

POLYESTER PRODUCTION USING DIFFERENT CARBON SUBSTRATES

Introduction of different carbon substrates to the culture medium of the microorganisms, at polymer production stage, can give way to the synthesis of tailor made products on will. The change introduced by different carbon source to the overall polymeric chain is expected to be reflected on the final product properties, such as crystallinity, biodegradation rate and other physicochemical features.

CALIBRATION

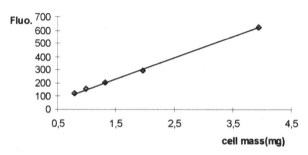

Figure 2. Calibration curve for stained *A.eutrophus* cells (grown on Fructose+4HBA).

Table 4. Polyester biosynthesis by A.eutrophus utilizing various co-substrates

C-Substrate (g)	Wet weight (g/L)	Dry cell weight (g)	Polymer recovered (g)	Yield (%)
Fru (20+10)	22.01	7.16	2.97	41.50
Fru+VL (18+2)	24.72	9.04	5.043 g	55.75
Fru+4HBA (18+2)	25.89	9.21	4.08	44.30
Fru+BrPPA (19+1)	22.84	7.90	2.38	30.16
Fru+BrPPA (18+2)	13.39	3.75	0.05	Trace

Fru: Fructose
VL: Valerolactone
4HBA: 4-hydroxybutyric acid
BrPPA: Bromopropionic acid

To achieve different polyester production, in addition to the major carbon source (ie. sucrose for *A.latus*, and fructose for *A.eutrophus*), co-substrates, like valerolactone, 4-hydroxybutyric acid, bromopropionic acid were added. The results were presented in Table 4 for *A.eutrophus* cells when fed on these co-substrates. The type of copolyester was determined by ^{13}C and ^{1}H-NMR (data not shown).

CONCLUSIONS

The results of the preliminary studies, indicate that the fluorimetric detection method developed is novel, fast, allowing on-line monitoring (that can be obtained from a dynamic culture), and is in good agreement with the widely accepted and applied conventional method (chloroform extraction). It is a mild procedure (as there is no sulphuric acid step), sensitive and reliable than conventional crotonic acid assay (an alternative spectrophotometric route). Thus, one may have the information about the percent yield of PHA, before hand, simply by using an aliquot from the culture and measuring its fluorescence. Additionally, it is possible to tailor the product by feeding with the desired carbon substrates.

REFERENCES

Howells E.,R., Single-cell protein and related technology., 1982., *Chem. Ind.*, 7:508

Lemoigne M., Products of dehydration and of polymerization of β-hydroxybutyric acid., *Bull. Soc. Chem. Biol.*, 1926., 8:770

Anderson A. J., and Dawes E. A., Occurrence, metabolism, metabolic role and industrial uses of poly hydroxyalkanoates, 1990, *Microbiol. Rev.*, 54(4): 450

Muller H. M., and Seebach, D., Poly(hydroxyalkanoates): a fifth class of physiologically important organic biopolymers, *Angew.* 1993, *Chem. Int. Ed. Engl.*, 32:477

Doi Y., *Microbial Polyesters,* 1990, VCH Publishers Inc., New York

Brandl H., Gross R. A.,. Lenz R. W., and Fuller R. C., Plastics from bacteria and for bacteria, 1990, in, *"Advances in Biochemical Engineering and Biotechnology"*, Ed. A. Fiechter, Springer Verlag, V. 41, pp78-93,

Lee S.Y.,, Bacterial polyhydroxyalkanoates., 1996, *Biotechnol. Bioeng.*, 49: 1

Pouton C. W., and Akhtar S., Biosynthetic PHV's and their potential in drug delivery., 1996, *Adv. Drug Deliv. Rev.*, 18:133

Steinbüchel A., and Valentin H.E., Diversity of bacterial PHA, 1995, *FEMS Microb. Letts.*, 128:219-228,.

Holmes P. A., Applications of PHB:a microbially produced thermoplastic, 1985, *Phys.Technol.*, 16:32

Shubert P., Steinbuchel A., and Schegel H. G., Cloning of the *A.eutrophus* genes for synthesis of PHB and synthesis of PHB in *E.coli*, 1988, *J. Bacteriol.*, 170(12):5837

Slater S. C., Voige W. H., and Dennis D. E., Cloning and expression of *E.coli* of the *A.eutrophus* H16 PHB biosynthetic pathway, 1988, *J. Bacteriol.,* 170(10):4431

Steinbuchel A., and Wiese S., A *Pseudomonas* strain accumulating polyesters of 3-hydroxybutyric acid and medium chain length 3-hydroxyalkanoic acids, 1992, *Appl. Microbiol. Biotechnol.,* 37:691

Timm A., and Steinbuchel, A., Formation of polyesters consisting of medium chain length 3-hydroxyalkanoic acids from gluconate by *P.aeruginosa* and other fluorescent *Pseudomonads.,* 1990, *Appl. Environ. Microbiol.,* 56(11):3360

Gürsel I., and Hasırcı V.,, Properties and drug release behaviour of poly (3-hydroxybutyric acid and various poly (3-hydroxybutyrate-co-hydroxyvalerate) copolymer microcapsules., 1995a, *J. Microencapsulation,* 12:185

Gürsel İ., and Hasırcı V., Microorganismal origin biopolymers., 1995b, *Science and Technique,* 334:97

Gürsel İ., Use of polyhydroxyalkanoates in the construction of biomedical drug release systems, 1995., *Ph. D. Thesis,* METU, Ankara, Turkey,

Yasin M., and Tighe, B.J., Polymers for biodegradable medical devices. VIII., 1992., *Biomaterials.,* 13:9

Holland S. J., Jolly A.M., Yasin M., and Tighe, B.J., Polymers for biodegradable medical devices. II., 1987, 8:289

Law H., J., Slepecky R. A., Assay of poly-β-hydroxybutyric acid., 1961, *J.Bacteriol.,* 82:33

Ward A. C., and Dawes E. A., A disk assay for poly-β-hydroxybutyrate., 1973, *Anal. Biochem.,* 52:607

Williamson D. H., and Wilkinson J. F., Isolation and estimation of the inclusions of Bacillus species., 1958, *J.Gen Microbiol.,* 19:198

Yamane T., Yield of poly-D(-)-3- hydroxybutyrate from various carbon sources: A theoretical study., 1993, *Biotechnol. Bioeng.,* 41:165

Suzuki T., Yamane T., and Shimizu S., Mass production of poly-β-hydroxybutyric acid by fed-batch culture with controlled carbon/nitrogen feeding., 1986, *Appl. Microbiol. Biotechnol.,* 24:370

Kim S. B., Lee S. C., Lee S. Y., Chang H. N., Chang Y. K., and Woo S. I., 1994., Production of poly(3-hydroxybutyric acid-co-hydroxyvaleric acid) by fed-batch culture of *Alcaligenes eutrophus* with substrate control using on-line glucose analyzer., *Enzyme Microb. Technol.,* 16:556

Linko S., Vaheri H., and Seppala J., Production of poly-β-hydroxybutyrate by *Alcaligenes eutrophus* on different carbon sources., 1993, *Appl. Microbiol. Biotechnol.,* 39:11

ANTIBIOTIC RELEASE FROM BIODEGRADABLE PHBV MICROPARTICLES

Dilek Şendil,[1] İhsan Gürsel,[1] Donald L. Wise,[2] and Vasıf Hasırcı [1]

[1]Middle East Technical University, Department of Biological Sciences, Biotechnology Research Unit, 06531, Ankara, TURKEY
[2]Department of Chemical Engineering and Center for Biotechnology Engineering, Northeastern University, Boston, MA 02115, USA

INTRODUCTION

The gingival tissue is constantly subjected to mechanical and bacterial aggressions. Gingivitis, which is the inflammation of the gingiva, is the most common form of gingival disease. The extension of the inflammation from the marginal gingiva into the supporting periodontal tissues marks the transition from gingivitis to periodontitis. Inflammation extends along collagen fiber bundles and goes through alveolar bone and may result in bone destruction[1]. Periodontitis is always preceded by gingivitis but not all gingivitis proceed to periodontitis.The selective removal or inhibition of pathogenic microbes with either systemic or topically applied antibacterial agents when combined with scaling and root planing is often an effective approach to treatment of specific disease active sites, severely advanced and/or refractory cases of periodontitis[2]. Tetracycline is the most abundantly tested and used antibiotic in the treatment of periodontal diseases. Clinical studies using tetracycline.HCl (TC) have shown it to have an effective spectrum of activity against many of the anaerobic microbes associated with the various periodontal diseases involving both adult and juvenile periodontitis patients. Tetracyclines have been most widely and successfully used in the treatment of juvenile periodontitis[3,4]. TC has several inherent properties which enhance its potential adjunctive use in the treatment of periodontal disease. These are the substantiality of TC to dentine and cementum surfaces, its ability to etch and/or remove the root surface smear layer and cause surface demineralization (chemical conditioning of the root surface), to delay pellicle and plaque formation, and to exhibit anti-collagenase activity. As the systemic use of antibiotics may cause several side effects (sensitivity, resistant strains, superinfections), the local administration of antibiotics has received considerable attention[5].

Recently, a novel biodegradable, biocompatible and biological origined polyhydroxyalkanoates started to draw attention from several fields of science, including medicine, pharmacy and cosmetics, and agriculture[6]. These polymers are ideal for use as biomedical materials due to their unique and interesting physicochemical features[7].

In this study, polyhydroxybutyrate-co-hydroxyvalerate (PHBV) with varying comonomer ratios were used in the construction of controlled antibiotic systems to deliver TC or neutralized TC (TCN) and the properties of the resultant systems were analyzed by in vitro release studies. The morphological changes before and after release were assessed by SEM studies. Also, to determine the extent of retention of the biological activity of the drug after microcapsule preparation procedures, bioassays were carried out using a tetracycline sensitive organism, E.Coli (C600) (MIC: 0.78).

PREPARATION OF TC AND TCN LOADED PHBV MICROPARTICLES

Two different forms of the drug was used during experiments (tetracycline.HCl and neutralized tetracycline, TC and TCN, respectively). PHBV (7, 14 and 22 % HV content) microcapsules were prepared with the TC by using double emulsion, solvent evaporation technique as described earlier[8,9]. Neutralization of tetracycline.HCl was done by mixing with equimolar NaOH solution and titrating to pH 7. This neutralized form was used in the preperation of microspheres with a slight modification of the above stated method in which drug was introduced into the PHBV solution in crystalline form without being dissolved in distilled water (i.e. single emulsion). Optimization of the encapsulation medium that will yield the highest microcapsule recovery with the highest drug encapsulation was carried out by changing the emulsifier and stabilizer concentrations (data not shown), and 2 % gelatin solution (100 mL) containing polyvinylalcohol (PVA) (1.0 mL, 4 % w/v) was found to yield the highest drug incorporation and microcapsule recovery.

Effect of polymer type on encapsulation efficiency, loading and release was investigated using PHBV copolymers with different HV contents (7, 14 and 22%) while keeping all other experimental conditions such as PVA, gelatin, and polymer concentration constant. Results showed that the highest encapsulation efficiency was obtained with PHB14. For the TC loaded microcapsules, it can be deduced that Table 1the encapsulation efficiency and loading of PHBV7 and 22 decreases, as the HV content is increased Gangrade et al observed a similar trend PHB, PHBV9, and PHBV24 where the drug loading values were 2.24, 1.82, and 1.93 %, respectively[10]. This could be explained with the lower crystallinity of PHBV22 microcapsules due to its higher valerate content compared to PHBV 7.

In case of TCN loaded microspheres, the encapsulation efficiency was slightly higher for PHBV22 than PHBV7, with PHBV14 behaving unpredictably as in our earlier studies (lower than the rest) (Table 1). It was observed that, during polymer drug mixing, TCN solubility was higher than TC and therefore mixing yielded a more homogeneous PHBV/TCN solution. This effect is thought to be more distinct in the case of PHBV22 polymer solution, because its dissolution was much more complete (increasing valerate content increases solubility). It thus forms a better drug dispersion that can hold more TCN

Table 1. Properties of TC and TCN loaded PHBV microcapsules at 25 ± 2 °C

Microcapsule Type	PVA (%,w/v)	Gelatin (%,w/v)	EE (%)		Loading*		Cum. Drug Released in 4 days(mg)	
			TC	TCN	TC	TCN	TC	TCN
PHBV7	1	2	30.1	63.3	5.0	11.0	0.43	0.22
PHBV14	1	2	35.9	51.9	6.7	9.3	0.43	0.41
PHBV22	1	2	25.8	65.3	4.9	11.6	0.68	0.22

*mg drug/100mg microcapsules

during microsphere formation. In TCN loading process, also the decrease in water solubility and introduction of the drug in powder form, prevented excessive drug loss into the aqueous phase and at the same time drug localization was much more on the polymeric coat than in hollow core. Higher TCN loading than. TC and on the other hand retarded release rates are an indication of these interactions between the drugs and the polymeric carrier.

In order to modulate release kinetics, several approaches are used. Among them are variation of the support material properties (such as using various forms of the polymeric carrier, as it is attempted in the present study) or modification of the drug itself (again as in this work changing of solubility). The release behaviours of two forms of the drug revealed that the TC has a much higher tendency to leach out than its counterpart TCN (Figures 1 and 2). This observation prompted us to co-encapsulate the two forms of the drugs (TC and TCN) in one preparation, and thus modify the overall release behaviour (Figure 3). To test this, TC and TCN (1:1 weight ratio) was co-encapsulated in PHBV7 microparticles. Encapsulation efficiency of this form was observed to be in between the separate forms, but closer to the TC preparation (Table 2). This might suggest that even a very small amount of aqueous solution within the microparticle decreases the homogeneity of the polymer-drug mixture and thus, microcapsule stability.

RELEASE STUDIES

Tetracycline.HCl loaded microcapsules (15 mg) or tetracycline loaded microspheres (15 mg) were dialysed in PBS (phosphate buffered saline; 100 mL) which was continuously stirred by a magnetic stirrer at ambient temperature. At certain time intervals, samples were removed (5.0 mL) from the release medium and tetracycline content was measured spectrophotometrically (at 360 nm). Following measurement, aliquots were returned. Results of triplicate tests were used to compute the released tetracycline. The data were plotted as amount of drug (mg) released vs time, and as cumulative release vs square root time. These plots were then used to determine; lag times, burst rates release constants.

It was observed that TC release from PHBV22 was about 90% in 100 h, while it was only 50% for PHBV7. The release with PHBV7, showed almost a zero-order behaviour and for PHBV22 this was closer to Higuchi prediction (see relevant plots in Figure 1). For PHBV14 at 100 h only 42% was released. The trend seen in loading and encapsulation efficiencies (Table 1) is also reflected on the release patterns for PHBV7, 14, and 22, which is contrary to the expectations that increasing HV content, increases loading and cumulative release for a particular drug). The unexpected PHBV14 performance was thought to be due to the difference in molecular weight of the polymer[10]. To check this PHBV14 was further characterized by viscosimetry, and NMR. Results indicated that the

Table 2. Properties of TC and TCN loaded microparticles*

Microcapsule Type	EE (%)	Loading*	Lag Time (h)	Slope ($h^{-1/2}$)
PHBV7+TC+TCN	36.17	6.8	0.689	0.0325
			5.035	
PHBV7+TC	26.37	5.0	0.406	0.106
			0.212	
PHBV7+TCN	63.9	11	0.535	0.025
			0.445	

*At 25 ± 2 °C
*mg drug / 100mg microparticles

polymer has considerably low molecular weight than the stated value by the producers (further confirmed by mechanical tests[11], (data not shown) possibly causing the deviation from the expected.

From the Higuchi plots (Mt/M_∞ vs $t^{1/2}$) of the results it is observable that in the case of PHBV22 the release yields an almost perfect fit ($R^2=0.9897$). Since this implies a behaviour like that of monolithic sphere, it looks as if either most of the drug is entrapped close to the surface (within the coat) of the PHBV22 type of microcapsules or diffusion of TC through the coat is easier possibly due to membrane discontinuity for this type. Possibly the copolymer coat entrapped fraction of drug is released giving a direct proportionality with square root of time which is obtained for PHBV7 and 14 cases only in the initial phase of the release. This was also supported by Gangrade and Price who found that a higher valerate content leads to a higher porosity and since the drug is also in the coat, the increased surface area due to the pore walls lead to higher release rates[10].

When Table 1 is examined it is seen that even though the results are very close, PHBV14 has the highest initial slope (most rapid release) followed by PHBV22 and PHBV7. Normally PHBV 22 should release the fastest due to its composition. The figures (Figure 1a and b) show that while at later stages PHBV22 maintains this rapid release, release from PHBV7 and 14 slows. It is known that PHV component of the copolymer leads to a more amorphous structure thus decreasing the crystallinity of the molecule. So, it might be possible that the drug could partition into this more amorphous, HV rich regions and is released through HV channels. The partition coefficient (0.22 o/w) of TC also suggests that the drug would prefer the more hydrated domains, like HV. The biphasic release observed in all cases (except PHBV22) in this study is nearly a characteristic of PHBV microparticles observed in many studies[12]. This result also suggests that in this short period of time (insufficient for the degradation of PHBV copolymer) release is mainly dependent on the diffusion of the drug through the pores of wall and thus diffusion is the rate determining step, and the initial phase of release was thought to occur mainly by dissolution and diffusion of drug entrapped close or at the surface of the microcapsules. The second and slower phase release was thought to involve the diffusion of drug entrapped within the inner part of the polymer matrix by means of aqueous channels of a network of pores[13]. This decline is also due to depletion of the drug content and because of the shape of the microcapsules, within which the surface facing the medium decreases as one goes towards the center from the surface. An interesting observation is the presence of a lag period indicative of the need for solvent penetration into the structure. It appears that this stage is quite significant in these microcapsules.

The original aim of neutralization of TC was to decrease its solubility and thus release rate. It also made possible the use of powder drug in o/w encapsulation. When the results (Table 2, Figure 2) are examined, it is observed that in PHBV7, the microspheres released ca. 20% in 800h. PHBV14 released ca. 55% and PHBV22 released only 18% in 480h. So the TCN looks mostly well entrapped especially within the microspheres of PHBV 7 and 22 for which encapsulation efficiencies are also quite high (Table 2). When the slopes are compared with those of TC released from same polymers, it is observed that the slopes with TCN are between 50% to 80% lower in the case of TCN. Thus much slower release is achieved with the use of TCN.

RELEASE OF TC AND TCN FROM PHBV7 MICROPARTICLES

Initial aim of using this approach was to obtain a bimodal release from the microsphere. As seen in Figure 3, the release behaviour is not bimodal but more in accordance with higuchi's expression. Likewise the value of the slope of the Higuchi curve for co-encapsulated TCN and TC is between the slopes of their individual 2 cases.

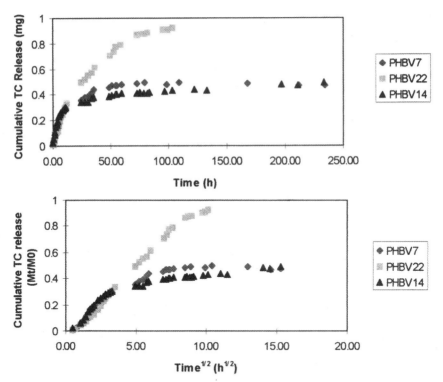

Figure 1. Release of TC from PHBV7, 14 and 22 microcapsules.

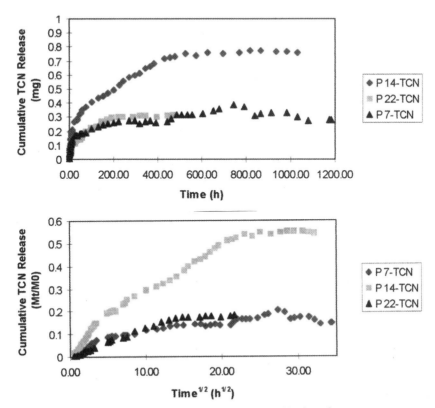

Figure 2. Release of TCN from PHBV7, 14, 22 microspheres.

Release plots showed that 45% of TC was released within ca. 300h for the two drug loaded case (Figure 3) but in 50h for TC microcapsules. Release of drug from TCN microspheres was ca.15% in 300h. This release data shows that two drug loaded microparticles give a release profile which is between separate loadings of TC and TCN, and closer to that of TC. The reason for this is that although it appears that there were two approaches, the loading process necessitated the encounter of organic PHBV solution and TCN thus leading to its dissolution. Therefore, encapsulation in powder form was not possible. The results confirm this process problem.

SCANNING ELECTRON MICROSCOPY (SEM) STUDIES

In order to investigate the morphological changes occurred on the TC and TCN loaded microcapsules, before and after release studies, following gold coating they were investigated by SEM. Photomicrographs obtained revealed that, the coat surface of the microcapsules were much porous at the end of the release period (Figure 4). This porosity was higher with PHBV22 microcarriers, indicating the possibility of surface erosion is another significant factor throughout the end of the release period.

CONCLUSIONS

In the present study, construction of a biodegradable drug release system for the treatment of periodeontal diseases was aimed. This was achieved by incorporating tetracycline (in two different forms) within PHBV microparticles. The parameters influencing the loading and the in vitro antibiotic release profiles (i.e. stabilizer and

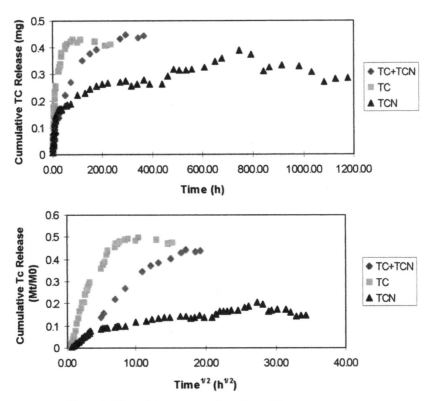

Figure 3. Effect of drug type on release from PHBV7 microcapsules.

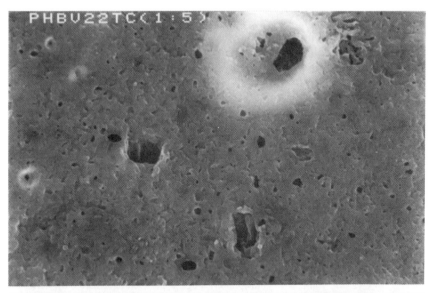

Figure 4.Scanning electron micrograph of TC loaded PHBV14 before (top) and after (bottom) release study.

emulsifier concentrations) were investigated. It was found that the encapsulation efficiency, drug loading and release rates were all closely related with the drug chemistry. When the neutralized form of the drug was used, encapsulation efficiency improved significantly, but a slower release rate than those obtained with the hydrochloride salt of the antibiotic was obtained. The data could not be interpreted as a zero-order behaviour but could be better described by release from spherical matrix system model (which depend on diffusion characteristics). SEM micrographs of the generated microparticles showed microporous, perfect spherical shapes that undergo surface erosion upon long term incubation in the release medium (ca. 50 days). Finally, the biodegradable release system developed has the potential for application as a local antibiotic release depot in the treatment of periodontitis, avoiding the problems such as the need for removal when the therapy is complete and the depot is depleted.

REFERENCES

1. Carranza, F.A., Odont, Jr., Glickman's *Clinical Periodontology*, 1990, W.B. Saunders Co,.Publ. New York
2. Drisko, C.L., Cobb, C.M., Killoy, W.J. et al., Evaluation of periodontal treatments using controlled-release tetracycline fibers: clinical response., 1995, *J Periodont.*, 66:692.
3. Smith, S.R., Foyle, D.M., Needleman, I.G., Pandya, N.V., The role of antibiotics in the treatment of periodontitis(part 1 -systemic delivery), *Eur. J. Prostbodont. Rest. Dent.*, 3(2), 79-86, 1994
4. Palmer, R.M., Watts, T.L.P., Wilson, R.F., A double-blind trial of tetracycline in the management of early onset periodontitis, *Journal of Clinical Periodontology*, 23, 670-674, 1996
5. Ünsal, E., Akkaya, M., Walsh, T.F., Influence of a single application of subgingival chlorhexidine gel or tetracycline paste on the clinical parameters of adult periodontitis patients., 1994, *J. Clin. Periodont.*, 21:351
6. Doi Y., *Microbial Polyesters.*, 1990, VCH Publishers Inc., New York
7. Holmes P.A., Applications of PHB: a microbially produced thermoplastic., 1985, Phys. Technol., 16:32
8. Gürsel, İ., Hasırcı, V., Properties and drug release behaviour of poly (3-hydroxybutyric acid) and various poly (3-hydroxybutyrate-hydroxyvalerate) copolymer microcapsules, 1995, *J. Microencapsulation*, 12:185,
9. Şendil, D., Antibiotic release from biodegradable microbial polyesters, 1997, *MSc. Thesis.*, METU, Ankara
10. Gangrade, N., Price, J.C., Poly(hydroxybutyrate-hydroxyvalerate) microspheres containing progesterone: preparation, morphology and release properties, *J. Microencapsulation*, 1991, 8:185
11. Gürsel, İ., Balçık, C., Arıca, Y., Akkuş, O., Hasırcı, V., and Akkaş N., Interpenetrating networks of polyhydroxybutyrate-co-hydroxyvalerate, and polyhydroxyethyl methacrylate : Mechanical properties., 1996, *Engineering Systems Design and Analysis*, 5:107,
12. Koosha, F., Muller, R.H., Davis, S.S., Polyhydroxybutyrate as a Drug Carrier, *Crit. Rev. Ther. Drug Carr. Sys*, 1989, 6:117
13. Akhtar, S., Pouton, W.C., Notarianni, J.L., The influence of crystalline morphology and copolymer composition on drug release from solution cast and melt-processed P(HB-HV) copolymer matrices, 1991, *J. Contr. Rel.*, 17:225

DRUG CARRIER SYSTEMS FOR BIOTECHNOLOGY DERIVED PRODUCTS

Filiz Öner

Hacettepe University
Faculty of Pharmacy
Pharmaceutical Biotechnology Department
Ankara, 06100, TURKEY

INTRODUCTION

Progress in biotechnology and increasing number of biotechnology derived drug products have led to the development of many new drug carrier systems.

First commercially available biotechnology derived recombinant DNA products are peptides and proteins but still there are problems to overcome. Proteins and peptides are easily degraded in the gastrointestinal tract and they can be administered exclusively by inconvenient parenteral routes. Alternative routes for peptide protein administration is becoming the major challenge in formulation development studies [1,2].

Recombinant vaccines take an important place among biotechnology derived pharmaceutical products. Subunit vaccines are highly purified antigens often produced by recombinant DNA technology and do not contain patogenic materials. Their safety profiles are much better than conventional vaccines but they elicit poor immune responces. In order to achieve long lasting antibody response with improved immunogenicity, effective adjuvants or adjuvant carriers need to be investigated [3-5].

Since the cloning of the first gene in seventies, there have been many advances in medicine at the molecular level of diseases. Gene therapy represents a new age for therapy of human diseases. Thus gene therapeutics may be the most promising topic of pharmaceutical biotechnology and suitable gene delivery systems are increasingly studied by various research groups [6-8].

BIOTECHNOLOGY DERIVED DRUG PRODUCTS

Today there are about 30 pharmaceutical biotechnology products on the market, and more than 400 products either in human clinical phase trials or at the health authorities for approval. Many biotechnology drugs are in development for diseases such as cancer, cystic fibrosis, AIDS, neurological disorders, headaches and infectious diseases.

Biotechnology derived drug products can be classified in four main groups as seen in Table 1. Effective carrier system requirements for peptides - proteins, vaccines and gene therapeutics are becoming more important due to their increasing therapeutic, diagnostic and prophylactic importance.

Table 1. Drug products derived from biotechnology

Peptides - Proteins	Vaccines	Gene Therapeutics	Small Biological Molecules
Recombinant DNA products	Recombinant Antigens	DNA, RNA	Antibiotics
Monoclonal Antibodies	DNA	Oligonucleotides	Vitamins
Animal Transgenesis Products		Antisense Oligonucleotides	Aminoacids

PEPTIDES - PROTEINS

Therapeutic peptides and proteins are becoming readily available through rapid advance in recombinant DNA technology. Hormones, growth factors, tissue plasminogen activators, interferons, interleukins, clotting factors and erytropoietins are the most known examples of these proteins and peptides [9,10]. Table 2 gives some of the products belonging to each group. In 1996, number of biotechnology drugs have increased 21% [11] when compared with 1995.

Table 2. Recombinant therapeutic proteins and peptides

Proteins - peptides	Products
Hormones	rh Insulin, rh Growth Hormone, rh Leutinizing Hormone, rh Follicle Stimulating Hormone
Growth Factors	EGF, CSF, rG-CSF, rGM-CSF,TGF, FGF, NGF, BDNF, PDGF, rhBMP-2, TNF, Fibronectin,Vitronectin, Neurotropin
Tissue Plasminogen Activators	rt-PA, second generation t-PA
Interferons	Interferon gamma-1b, rInterferon alfa-2b, rInterferon alfa-2a, rInterferon, Interferon alfa-n3, rInterferon beta-1a
Interleukins	Interleukin-2, rhInterleukin-3, rhInterleukin-4, rhInterleukin-6, rh IL-10,rh il-12, glycosylated rhIL-6
Clotting factors	rAHF, rFVIII, rhFIX, rFVIIa
Erythropoietins	Epoetin alfa, trombopoietin

EGF: epidermal growth factor, CSF: colony stimulating factor, G-CSF granulosyte colony stimulating factor, GM-CSF: granulocyte macrophage colony stimulating factor, BNDF: brain derived neurotropic factor , TGF: transforming growth factor, FGF: fibroblast growth factor, NGF: nerve growth factor, PDGF: platelet derived growth factor, BMP: bone morphogenic protein, TNF: tumor necrosis factor, ILs: interleukins,AHF: antihemophilic factor, t-PA: tissue plasminogen activator, FVIII: factor VIII, FIX: factor IX, FVIIa: factor VIIa, r:recombinant

Monoclonal antibodies (MoAbs) can be produced by hybridoma technology or by recombinant DNA technology.The first therapeutic product of MoAb technology, immunosupresant Muromonab - CD3, is used to prevent rejection of transplanted kidney[12]. Many other monoclonal antibodies are in development against diseases such as cancer and infectious diseases [13-15]. The use of some monoclonal antibodies are seen in Table 3.

Large animal transgenesis is another way of producing peptides and proteins. Mammary glands of large animals can be used as bioreactors for proteins that can be purified in high amounts from milk [16-17]. Human proteins under development in milk are α1 antitrypsin, anti-thrombin III, collagen, Factor IX, Factor VIII, fibrinogen, human hemoglobin, lactoferrin and tissue plasminogen activator, as listed in Table 4.

Table 3. Monoclonal antibodies

Product	Use
Muromonab -CD3	Prevention of renal transplant rejection
Anti-fibrin antibody	Blood clot imaging agent
Human anti-hepatitis B antibody	Liver transplant patients with chronic active HepatitisB
Mab 14, 18, Mab 14G2a, Mab COL-1	Neuroblastoma, solid tumors, gastric, colorectal cancer
BR96-Doxorubisin conjugate	Cancer
Anti-IgE humanized Mab	Asthma
E5 Mab	Gram negative sepsis
AD-519 Mab, Anti HIV to CD4 region of gp120 protein	HIV infection, AIDS
Iodine 131 -intact IgG	Colorectal cancer, non Hodgkin's B-cell lymphoma
Tecnetium-99m-Fab'fragment	Extent of disease staging of cancers

Table 4. Human proteins under development in milk of large transgenic animals

Protein	Transgenic species	Use
Alfa-1 anti-protease inhibitor	Goat	Alfa -1 antitripsin deficiency
Anti-trombin III	Goat	Sepsis
Collagen	Cow	Burns, bone fractures
Factor IX, VIII	Sheep.pig	Hemophilia
Hemoglobin	Pig	Blood substitute
Fertility hormones	Goat, cow	Infertility, contraceptive vaccines
Human serum albumin	Goat	Burns, surgery,shock
Lactoferrin	Cow	Gastrointestinal infection
Protein C	Pig, sheep	Protein C deficiency
Tissue plasminogen activator	Goat	Heart attacks, trombosis, pulmonary embolism
Alfa-1 antitripsin	Goat	Anti-inflammatory
Fibrinogen	Sheep, pig	Surgery. burns
Monoclonal antibodies	Goat	Anti-colon cancer

Carrier Systems

The potential therapeutic applications of these increasing number of biotechnology derived substances mostly depend on their successful delivery to desired sites and organs by suitable routes.

The physicochemical properties of peptides and proteins make their formulation difficult. Proteins and peptides are large, easily degraded, easily metabolized molecules and they can not be absorbed easily through the mucosal areas.

Biotechnology derived peptide-protein drugs are available almost in the parenteral dosage form, only a few of them are applied via mucosal routes and parenteral forms of peptides and proteins have some problems need to be overcome. These problems are regarded with the formulation, processing, sterilization and administration of peptides and proteins. Alternative delivery routes such as oral, nasal, buccal, rectal, vaginal, dermal, transdermal and pulmonary are increasingly investigated.

Oral Delivery

Oral administration is the most traditional route and it is always the first choice of the patients. Administering a drug by oral route, blood levels can be controlled, side effects can be reduced and patient compliance is improved, however oral delivery of peptides and proteins present some problems. Gasrointestinal proteases inactivate these molecules, first pass effect cause inactivation of peptides and proteins due to hepatic enzymes and mucosal absorption of them is limited because of their large molecular size.

There are some solutions to improve the oral delivery of proteins and peptides. These can be summarized as follows ;
-Protection of the molecule by enteric coating
-Protection of the molecule by enzyme inhibitors
-Increasing the absorption by enhancers (surfactants, bile salts)
-Improving the molecular uptake with lipid carriers

Delivery of peptides and proteins in polymeric carriers include biodegradable polymer matrix systems [18-19] (microspheres, nonospheres, microbeads), diffusion controlled and enteric coated porous membranes (microcapsules, nanocapsules), osmotic pressure controlled systems and hydrogels[20-21] Lipid carriers for peptide protein drugs are emulsions, liposomes, micelles, lipoproteins and lipospheres [22-26] Cell carriers such as erythrocytes and erythrocyte ghosts are another group under investigation for peptides and proteins [27]

Oral Emulsions for Recombinant Human Insulin. Among the strategies which are proposed to improve the oral administration of peptides the potential of emulsions has been investigated [24-25]. In a study we have tested oral W/O/W emulsions for recombinant human insulin. Emulsions prepared by one step and two step emulsification procedures which contained aprotinin as protease inhibitor. In vitro release studies showed that incorporating insulin in the inner aqueous phase of the emulsions prepared with two step procedure protected the drug from environment and the drug was further protected with aprotinin in the outher phase as given in Figure1[28]. A hypoglycemic effect was observed after oral administration to the normal rabbits as seen in Figure 2 [29].

Topical Carriers for Recombinant Proteins in Wound Healing

Recombinant growth factors can be applied locally, for instance to promote wound healing. Wound healing effectiveness of recombinant growth factors (recombinant GM-CSF and EGF) are enhanced by using different carrier systems in our studies[30-32]

O/W emulsion, niosome and liquid crystalline carriers for rGM-CSF are evaluated with a mouse model. Wounds were considered healed when the moist granulation was no longer visible and wound was covered with a continuous layer of epitelium. Besides the measurements of changes in wound area , immunohistochemical and histological evaluations were done according to presence of crust, epidermal rejeneration, presence of acute inflammatory elements and collagenation of granulation tissue. Photographs of the biopsies stained with hematoxylin-eosin are seen in Figure 4 - 7. rh GM-CSF containing groups displayed more effective wound healing due to the neutrophil activating and chemotactic effects of active substance. Base vehicles also displayed wound healing effects due to their occlusive and repairing properties. However liquid crystalline vehicle could not display similar healing effectiveness because of its high surfactant content. All groups were found to be significantly different from the control group indicating that all vehicles can display wound healing effects to a certain extent.

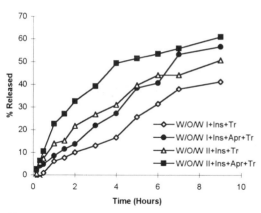

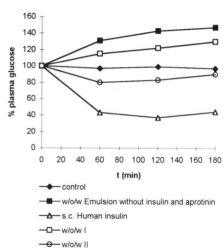

Figure 1. In vitro release profiles of recombinant from W/O/W emulsions in simulated intestinal fluid (from ref.29). W/O/W I : emulsion prepared with one step method, W/O/W II:emulsion prepared with two step method, Ins: Insulin, Tr: Tripsin.

Figure 2. Plasma glucose levels after single dose of oral human insulin from W/O/W emulsions (from ref. 29).

Emulsion and niosome vehicles are found as effective wound healing agents and wound healing effectiveness of rh GM-CSF can be potentialized by the use of proper carrier systems.

Chitosan and perflorocarbon gels for EGF in wound healing were evaluated with a rabbit model. In all experiments carrier formulations with growth factors showed enhanced healing effects when compared to their aqueous solution forms [31,32.]

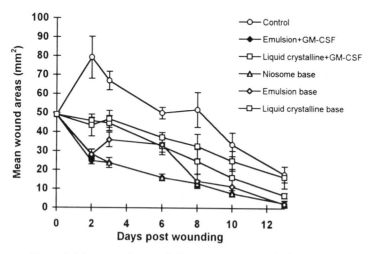

Figure 3. Mean wound areas of all groups versus control (from ref.30).

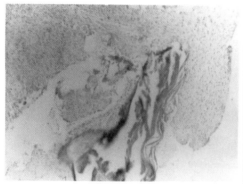

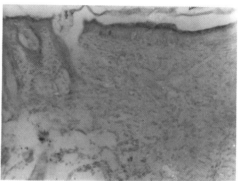

Figure 4. Photomicrograph of control group biopsy (from ref. 30).

Figure 5. Photomicrograph of biopsy from emulsion group containing rh GM-CSF (from ref. 30).

VACCINES

Vaccines are the second largest category of biotechnology products. They are being developed to prevent infectious diseases such as hepatitis B, influenza, pertussis, herpes simplex. Therapeutic vaccines are also developed to cure AIDS, rheumatoid arthritis, cancer, multiple sclerosis, peptic ulcers, psoriasis and others. Conventional vaccines are made of live attenuated or killed organisms. Unlike conventional vaccines new generation vaccines may contain recombinant DNA derived antigens, synthetic peptide antigens or DNA. Many of these new vaccines elicit poor immune responses and they need help in the form of an adjuvant.

Vaccine Adjuvants

Adjuvants are used to increase the humoral or cellular immune response to an antigen. An adjuvant can act with several mechanisms. It acts as a depot to retain high concentration of the antigen at the injection site. It is believed that sustained release of the antigen has a positive effect on the immune response. Adjuvants can create an inflammatory response and by attracting the macrophages and T-lymhocytes to the antigen source they can stimulate cytokine release can generate antibody responses.

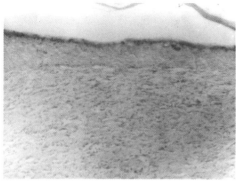

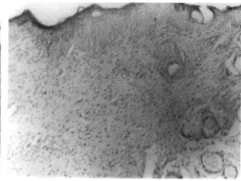

Figure 6. Photomicrograph of biopsy from niosome group containing rhGM-CSF (from ref. 30).

Figure 7. Photomicrograph of biopsy from liquid containing rhGm-CSF (from ref. 30).

Adjuvant formulations have two components as being small molecule immunomodulators and two phase disperse systems. Most known examples of the first group are muramyl dipeptide (MDP), monophosphoryl lipid A(MPL) and quillaria saponaria molina saponin [33-36]

In the second group there are emulsions, suspensions, niosomes, liposomes and iscoms[37-41]. Aluminum salts or alum is the most traditional adjuvant and it is the only formulation that can be used in human[35]. Freund's incomplete and complete adjuvants are mineral oil emulsions and they can be used only in animal experiments [5.]

The new hepatitis B virus vaccines consist of hepatitis B virus surface antigen (HBsAg) adsorbed to alum[42]. In many developing and developed countries Hepatitis B virus carriers are increasing and effective immunization is an important requirement. For HBV immunizations there has been a search to find an adequate adjuvant which can potentate the immune response with reduced amount of the antigen, with minimal side effects and without second or third booster injections.

In our laboratories we have tested adjuvant carriers for hepatitis B surface antigen. Water-oil-water multiple emulsion, niosome and aluminum phosphate carriers were compared with commercial vaccine by using mouse model. Photomicrographs of the emulsion and niosome systems are seen in Figure 8.

Toxicity tests are important for evaluating adjuvants and there are five recommended tests in the literature [43].
These include;
-Cytotoxicity assay in cultured monolayers
-Creatine phosphokinase activity in the serum
-Weight gain test
-Intracutaneous toxicity test
-Systemic toxicity and footpad edema

Among various available toxicity tests creatine phosphokinase and weight gain tests are used in the study. According to the creatine phosphokinase test results all of the adjuvants we have prepared are acceptable (acceptable creatine phophokinase activity levels are under 2000 IU[5]) as seen in Table 5. Weight gain test is still reliable measurement and if there is endotoxin present, weight of a mouse decrease during 24 hours[43]. In our study with weight gain measurements we have obtained acceptable results[44] which are given in Figure 9. On day 44 delayed type hypersensitivity was tested by observing skin reactions there were no reaction observed.

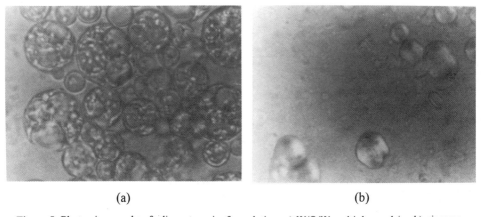

(a)   (b)

Figure 8. Photomicrographs of adjuvant carrier formulations a) W/O/W multiple emulsion b) niosome.

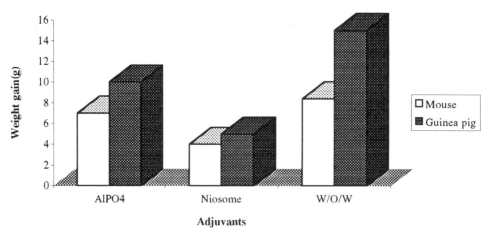

Figure 9. Weight gain measurements of mice and guinea pigs immunized with niosome, W/O/W emulsion and aluminiumphosphate.

GENE THERAPEUTICS

Recently use of genes as therapeutic agents show great potential for the treatment of inherited or acquired diseases. Delivery of foreign DNA into eucaryotic cells and expression of it, is becoming increasingly important and efficient carrier systems are required. As seen in Table 6 viral vectors are adenoviruses and retroviruses and besides their superior

Table 5. Creatine phosphokinase toxicity test

Formulations	Creatine phosphokinase activity
W/O/W emulsion	935
Niosome	182
Aluminum phosphate	423
Commercial vaccine	321

Table 6. Advantages and disadvantages of gene delivery systems

System	Advantages	Disadvantages
Retrovirus	Small genome High transfer efficiency No toxicity to host	Potentially mutagenic Limited insert capacity (10-15 kb)
Adenovirus	High infection efficiency High virus titers	Complicated vector genome May cause immune reactions Limited insert capacity
Cationic lipids	No infectious reactions No DNA size limit No toxicity	Poor targeting Low transfection efficiency Difficult in vivo administration
Receptor Mediated Delivery systems	No infectious reactions Possible specific targeting No DNA size limit	Low transfection efficiency Difficult in vivo administration Possible immunogenicity

Table 7. Gene therapeutics under development

Gene therapeutic	Use	Clinical Stage
Cystic fibrosis trnsmembrane protein (CFTR) adenoviral vector	Cystic fibrosis	Phase I/II
HS-tk(herpes simplex timidin kinase)	Brain cancer	Phase I/II/III
RV-AS-KRAS Ad-p53 RV-p53	Non-small Lung Cancer	Phase I
Retrovirus cellular gene therapy Anti-HIV T-cell therapy	HIV infection	Phase I/II
IL-2-gene modified tumor	Breast cancer	Phase I
Cytosine deaminase gene- adenoviral vector	Colon cancer	Phase I

transduction properties they have important disadvantages [45]. One of the major challenge in gene therapy is to develop safe and efficient synthetic non viral gene delivery systems that protects the DNA or RNA from nuclease degradation and deliver them to the target cells. Side effects due to the increased translation or transcription have to be minimized by appropriate targeting. DNA have to be targeted to nucleus and RNA have to be targeted to cytoplasm in order to be expressed. Thus, large and highly charged DNA and RNA have to cross several barriers before reaching to the target.

There are around 30 gene therapeutics under development in phase I; II or III clinical studies. Some examples are given in Table 7. These are primarily designed against cancers, HIV infections and cyctic fibrosis and most of them are viral systems[47].

Non-viral Gene Delivery Systems

Cationic liposomes are one of the first approaches for in vivo gene delivery with cationic lipids and are applied successfully . However plasma DNAses are interfere with the them. DNA-lipid complexes are reported to be more effective than conventional liposomes for gene delivery[47]. Cationic lipids are able to load high amounts of DNA by complexing it with electrostatic interactions. These systems are efficient nonviral transfection agents for animal cells in vitro.

Oil in water emulsion-DNA complexes are another alternative gene delivery systems and surfactant properties are found effective on the transfection characteristics of the systems[48].

Polymeric gene delivery systems are also designed to enhance stability and cellular uptake of DNA, DNA-gelatin nanospheres efficiently transfected transformed lung epithelial cell and deliver CFTR gene to rabbit lungs for the treatment of cystic fibrosis[49].

Receptor - mediated gene delivery can be achieved by using DNA - protein complexes.Transferrin - polylysine - DNA conjugates represent one of the major vector systems used in vitro transfection[5].These complexes can interact with receptors on the target cells and DNA internalized into the cell by receptor - mediated endocytosis.

Novel fusogenic peptide - DNA complexes are reported as a promising way of achieving efficient gene delivery either by non-specific or receptor mediated mechanisms[46].

CONCLUSIONS

Recent studies demonstrate that there are possibilities for the applications of new drug carrier systems to the delivery of recombinant peptides-proteins, vaccines, and genes. The challenge of pharmaceutical scientists is the production of stable, safe and effective carriers for these new biological molecules. Further studies are necessary to accelerate the availability of many new biotechnology derived therapeutic molecules.

REFERENCES

1. A. P. Sayani, and Y.W. Chien, 1996, Systemic delivery of peptides and proteins across absorptive mucosa, in: *Critical Reviews in Therapeutic Drug Carrier Systems* (S. Bruck,ed.),**13**(1,2):85-184.
2. P. Couvreur, and F. Puisieux, 1993, Nano- and microparticles for the delivery of polypeptides and proteins , *Advanced Drug Delivery Reviews*, **10**: 141-142.
3. D.Kline, J. Hares, R. Langer, 1996, Adjuvant- active polymeric microparticulate vaccine-delivery systems, in: *Microparticulate Systems for the Delivery of Proteins and Vaccines*, (S. Cohen, and H. Bernstein, eds.),pp. 349- 380, Marcel Dekker Inc, New York.
4. A.C. Allison, 1989, Antigens and adjuvants for a new generation of vaccines, in: *Immunological Adjuvants and Vaccines*, (G. Gregoriadis, A.C. Allison, and G. Poste, eds.), pp. 1-12, Plenum Press, New York.
5. D.E.S. Stewart-Tull, 1989, Recommendations for the assessment of adjuvants (Immunopotentiators), *in: Immunological Adjuvants and Vaccines*, (G. Gregoriadis, A.C. Allison,and G. Poste), pp. 213-226, Plenum Press, New York.
6. J.Y. Legendre, and F.C. Szoka,Jr, 1992, Delivery of plasmid DNA into mammalian cell lines using pH- sensitive liposomes: comparison with cationic liposomes, *Pharm. Res*, **9**:1235- 1242.
7. E. Wagner, M. Colten, R. Foisner, and M.L. Birnstiel, 1991, Transferrin-polycation-DNA complexes: the effect of polycations on the structure of the complex and DNA delivery to cells, *Proc. Natl. Acad. Sci. USA*, **88**:4255- 4259.
8. R.J. Mumper, M.K. Barron, K. Anwer, R. L. Lessard, Q. Liu, H. Nitta, H. Alila, and A. P. Rolland,1995, Interactive polymeric gene delivery systems for enhanced muscle expression, *Pharm. Res*, **12**, 80.
9. J. Buchanan, 1992, Recombinant DNA derived pharmaceuticals,in: *Recent Advances in Pharmaceutical and Industrial Biotechnology*, (A.A.Hıncal,and H.S.Kaş, eds.), pp 59-70 , Editions de Santé, Paris.
10. T.J. Maher,and K.M. Hull, 1992, Hormone production in biotechnology,in: *Recent Advances in Pharmaceutical and Industrial Biotechnology*,(A.A. Hıncal, and H.S.Kaş,eds.) pp.110-120 , Editions de Santé, Paris
11. Biotechnology therapeutic medicines and vaccines under development,1996, *Genetic Engineering News* (PhRMA Survey 1996), **15**:29-34.
12. F.M. Brodsky,1988, Monoclonal antibodies as magic bullets, *Pharm. Res*, **5**(1):1-9.
13. T.A. Waldmann,1991, Monoclonal antibodies in diagnosis and therapy, *Science*, **252**:1657-1662.
14. B.A. Khaw, and J. Narula,1992, Advances in the in vivo diagnostic applications of monoclonal antibodies in cardiovascular diseases,in: *Recent Advances in Pharmaceutical and Industrial Biotechnology*, (A.A. Hıncal,and H.S. Kaş, eds.), pp.82-98, Editions de Santé, Paris.
15. D. Brixner,1991, Monoclonal antibodies: From the research lab to the pharmacy, *Pharm. Practice*, **4**(2): 94-102.
16. I. Wilmut,A. L. Archibald, M. Mc Clenghan, J.P. Simons, C.B.A. Whitelaw and A.J. Clark, 1991, Production of pharmaceutical proteins in milk, *Experientia*, **47**:905-912.
17. J. Denman, M. Hayes, C. O'day, T. Edmunds, C. Barlett, S. Hirani, K.M. Ebert, K. Gordon, and J. M. McPherson, 1991,Transgenic expression of a variant of human tissue-

type plasminojen activator in goat milk, purification and characterization of the recombinant enzymes, *Bio/Technology*, 9:839-843.

18. S. Cohen, T. Yoshioka, M. Lucarelli, L.H. Hwang, and R. Langer,1991,Controlled delivery systems for proteins based on poly (lactic/glycolic acid) microspheres, *Pharm. Res*, **8(6)**:713-720.

19. J. Kreuter,1991, Nanoparticle based drug delivery systems, *J. Control. Rel*, **16**: 169.

20. P.J. Blackshear, 1987, Implantable pumps for insulin delivery: current clinical status, *in Drug Delivery Systems*,(P. Johnson,and J.G. Lloyd-Jones, eds.) pp.139-149, Ellis Horwood Ltd.,Chicester.

21. K. Park,W.S.W. Shalaby, and H. Park, 1993, Controlled release drug delivery systems, in: *Biodegradable Hydrogels for Delivery Systems*, (K. Park,W.S.W. Shalaby, and H. Park, eds.), pp.1-13, Technomic Publishing AG, Basel.

22. T.Trenktrog, B.W. Muller, and J. Seifert,1995, In Vitro-investigation into the enhancement of intestinal peptide absorption by emulsion systems, *Eur. J. Pharm. Biopharm*, **41**(5): 284-290.

23. G. Gregoriadis,1995, Engineering liposomes for drug delivery: progress and problems, *TIB TECH*, **13**: 527-537.

24. W.A. Ritschel, 1990, Gastrointestinal absorption of peptides using microemulsions as delivery systems, *Gattefosse*, **83**:7-22.

25. A.T.Florence, and G.W. Halbert,1991, Lipoproteins and microemulsions as carriers of therapeutic and chemical agents, in: *Lipoproteins as Carriers of Pharmaceutical Agents*, (J.M. Shaw.ed.), pp.141-174, Marcel Dekker Inc, New York.

26. S. Amselem, C.R. Alving, and A.J. Domb, 1996, Lipospheres for vaccine delivery,*in, Microparticulate Systems for the Delivery of Proteins and Vaccines ,(* S.Cohen,and H. Bernstein,eds.*)* pp. 149- 169, Marcel Decker Inc. New York.

27. M-I. Garın, R-M. Lopez, S. Sanz, M. Pinilla,and J. Lıque, 1996, Erythrocytes as carriers for recombinant human erythropoietin, *Pharm Res*, **13**(6):869-874.

28. H. Hudayberdiyev, F. Öner, and A.A. Hıncal, 1997, Oral multiple W/O/W emulsion formulations for recombinant human insulin, *Abst. J. Control. Rel*, **48**: 346-347.

29. F. Öner, H. Hudayberdiyev, İ. Ünsal and ,A.A. Hıncal, 1997, Oral delivery of recombinant insulin by water/ oil/water emulsion, *4th Int. Biomedical Sci Tech.Symp*, pp. 85-86.

30. E. Memişoğlu, F. Öner, A. Ayhan, İ. Başaran, and A.A. Hıncal, 1997, In vivo evaluation of rh GM-CSF wound healing efficacy in topical vehicles, *Pharm. Dev. Tech*, **2**:171-181.

31. E. Memişoğlu, F. Öner, H.S. Kaş, L. Zarif, A. Ayhan, İ. Başaran, and A. A. Hıncal, 1997, EGF wound healing in florocarbon and chitosan gels in a rabbit model, *4th Int.Biomedical Sci.Tech. Symp*, pp ..37-38.

32. E.Memişoğlu, F. Öner, L. Zarif, A. Ayhan, İ. Başaran, H.S. Kaş, and A.A. Hıncal,1997, Florocarbon based systems: wound healing potency in rabbit model with epidermal growth factor, *4th Int.Biomedical Sci. Tech: Symp*, pp.103-104.

33.M.E. Powell,1996, Drug delivery issues in vaccine development, *Pharm.Res*, **13** (12), 1777-1785.

34. C.R. Kensil, 1996, Saponins as vaccine adjuvants, *Critical Reviews in Therapeutic Drug Carrier Systems*, **13** (1,2):1-55.

35. E.T. Rietschel, L.Brade, and U. Schade, 1989, Bacterial endotoxins: relationships between chemical structure and biological activity in: *Immunological Adjuvants and Vaccines*, (G.Gregoriadis, A.C. Allison, and G.Poste, eds.), pp.61-74 ,Plenum Press, New York.

36. F. Ellouz, A. Adam, R. Ciorbaru, and E. Lederer, 1974, Minimal structural requirements for adjuvant activity of bacterial peptidoglycan derivatives, *Biochemical-Biophysical Research Communications*, **59** (4): 1317-1325.

37. M.Z. Khan, J..P.Opdebeech, and I.G.Tucker, 1994, Immunopotentiation and deliverysystems for antigens for single step immunization: recent trends and progress, *Pharm. Res*, **11** (1): 2-11.

38. R. Bomford, 1989, Aluminium salts, perspectives in their use as adjuvants,in: *Immunological Adjuvants and Vaccines*, (G. Gregoriadis, A.C. Allison, and G. Poste, eds.), pp. 35-43, Plenum Press, New York.

39. M. Brewer, J.Alexander, 1992, The adjuvant activity of nonionic surfactant vesicles (niosomes) on the Balb/c humoral response to bovine serum albumin, *Immunology*,**75**: 570-575.

40. A.A. Allison, and G. Gregoriadis, 1974, Liposomes as immunological adjuvants, *Nature*, **252**, 252.

41. B. Morein,B. Sundquist,S. Höglung, K. Dalsgaard, and A. Osterhaus, 1984, Iscom,A novel structure for antigenic presentation of membrane proteins from enveloped viruses, *Nature*, **308** (29):457-460.

42. N.E. Byars, G. Nakano, M. Welch, and A.C. Allison,1989, Use of syntex adjuvant formulation to augment humoral responces to hepatitis B virus surface antigen and to influenza virus hemaglutinin, in: *Immunological Adjuvants and Vaccines*, (G.Gregoriadis, A.C.Allison, and G. Poste,eds.), pp.145-152, Plenum Press, New York.

43. D.E.S. Stewart-Tull, 1996, Toxicity testing of adjuvants, *Proceed. 8th. Int. Pharm.Technol. Symp*, pp. 48-49.

44. A. Eratalay, and F. Öner, 1997, Unpublished results.

45. C. Coutelle, 1996, Gene therapy for inherited genetic disease: possibilities and problems, in : *Targeting of drugs, Strategies for Oligonucleotide and Gene Therapy*, (G. Gregoriadis, and B. McCormack,eds.), pp.1-13,Plenum press, New York.

46. A. Rolland, 1996, Controllable gene therapy: recent advances in non viral-gene delivery, in: *Targeting of Drugs, Strategies for Oligonucleotide and Gene Therapy*, (G. Gregoriadis, and B. McCormack,eds.), pp.79-97, Plenum Press, New York.

47. F.D. Ledley, 1996, Pharmaceutical approach to somatic gene therapy, *Pharm Res*, **13** (11): 1595-1614.

48. F. Liu, J.Yang, L. Huang, D. Liu, 1996, Effect of non-ionic surfactants on the formation of DNA/emulsion complexes and emulsion mediated gene transfer, *Pharm. Res,***13** (11):1642-1646.

49. S.M. Walsh, T.R. Flotte, V. L. Troung -Le, R. Rubenstein, S. Beck, T. August, P. Zeitlin, and K.W. Leong, 1997, Combination of drug and gene delivery by gelatin nanospheres for the treatment of cystic fibrosis, *Proceed Int.1. Symp. Control. Rel. Bioact. Mater*, **24**:75-76.

TISSUE ENGINEERING OF LIVER

Y. Murat Elçin

Ankara University
Science Faculty
Chemistry Department
Tandoğan, 06100 Ankara
Turkey

INTRODUCTION

Transplantation of a replacement part obtained from the patient himself, a living relative, a human cadaver, or a donor animal have been a way of therapy to compensate the deficient organ mass or organ function for more than forty years. Fulminant hepatic failure has an exceedingly high mortality. Liver transplantation is the treatment option of choice, and has been a successful alternative for patients with severe liver disease. However, its high cost, complexity of the procedure, and the scarcity of donor livers have been the main limitations of this process[1,2].

Liver has important role in a broad spectrum of functions, such as synthesis, metabolism, storage and release of several biochemical substances, detoxification, inactivation and several others[3]. That's why the early detoxification systems which were basically extensions of technology developed for chronic renal failure had very limited successes. Attempts to remove or dilute putative toxins in the blood have been unsuccessful in improving survival rates. The use of biocompatible interfaces with blood or plasma and current hepatocyte culture techniques have led to the development of novel support systems[4].

One of the most promising approaches for restoring liver function involves the use of cultured hepatocytes that will be part of an extracorporeal device. Some of these extracorporeal systems have been successful in sustaining patients with acute liver failure, preparing patients for liver transplantation, improving survival and quality of life for patients for whom liver transplantation is not a therapeutic option[5-7].

Transplantation of isolated hepatocytes to supplement deficient or failing liver function is a novel technique that offers several significant applications including the whole liver replacement therapy[8] (Figure 1). Hepatocyte transplantation may replace the hepatic function in patients with fulminant hepatic failure, hepatic dysfunction or inborn errors of metabolism[8,9]. Hepatocytes are anchorage-dependent, so cell-cell and cell-matrix interaction plays an important role in the success of maintaining hepatic function from transplanted cells.

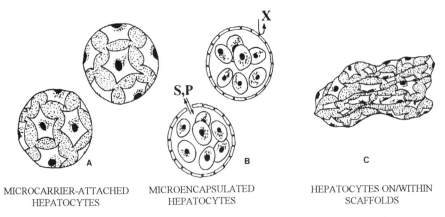

MICROCARRIER-ATTACHED HEPATOCYTES	MICROENCAPSULATED HEPATOCYTES	HEPATOCYTES ON/WITHIN SCAFFOLDS

Figure 1. Routes of hepatocyte transplantation other than direct injection. A: attachment of hepatocytes onto porous microcarriers [immunosuppression is needed]; B: microencapsulation of hepatocytes within spherical, ultrathin and semipermeable membranes [permeable to substrates (S) and products (P) of the cell metabolism, impermeable to hepatocytes, lymphocytes, antibodies and bacteria (X); immunosuppression is not needed]; c: attachment of hepatocytes onto/within porous biodegradable polymeric scaffolds [immunosuppression may be needed].

Promising preliminary studies using primary hepatocytes and polymeric biomaterials have demonstrated at least partial replacement of normal biological function using a cell transplantation approach[10].

These specifically engineered biomaterials have been utilized as a means of support for anchorage dependent hepatocytes and successful function of the transplanted cells have been documented in animal models, by means of albumin, urea and other liver-specific markers' production, and clearance of bilirubin and urea metabolites[11-13]. By using functional healthy cells with extracellular components that are *in vitro* cultured inside or on such supports can serve well for hepatocyte transplantation. Implantable tissue engineered bioartificial liver is a potentially important approach that can provide permanent hepatic support, provided that it is properly integrated into the body. The aim of this review is to make an overview on the developments of implantable systems, in the last decade.

HEPATOCYTE INJECTION

Hepatocyte transplantation techniques are suddenly becoming popular options to liver transplantations in a wide range of experimental contexts. However, there have been attempts to transplant hepatocytes or hepatic tissues in mammals going back for more than two decades[14,15]. Scientists have faced with difficulties and inconsistencies related to hepatic support systems with transplanted free hepatocytes through direct injection into various parts of the body[16,17]. Temporary but promising results have been reported in hepatocyte transplantation studies in several animal models of liver disease[18,19]. Stromal tissue was the extracellular matrix provider in direct injections that was needed for growth and differentiated function of the hepatocytes.

Liver has been proposed to be an optimal site for transplantation of hepatocytes, but the formation of cell aggregates in distal portal branches, sinusoids, and central veins have caused significant necrosis, as well as severe hypertension, both pulmonary and portal[20].

Spleen has also been chosen as a transplantation site for hepatocytes and improved long-term survival with differentiated hepatic function was observed upto two years after cell injection therapies[21]. The main limitations of spleen as a transplantation site were reported as the insufficient area for required injection volume, complications distinquished between transplanted and host hepatocytes, and limited proliferative capacity of the cell aggregates[21].

An alternative way of hepatocyte transplantation is the intraperitoneal route which allows higher quantities of cells to be injected with minimal trauma to the recipient. Encouraging findings have been reported by the use of peritoneum as transplantation site[22,23]. Lungs, portal vein, renal capsule, subcutaneous tissue, as well as intracapsular dorsal fat pads have also been evaluated as hepatocyte transplantation sites with varying success[24]. Immunosuppression was needed in direct injection routes to prevent the rejection of hepatocytes by the host[19]. In general, the control of placement, attachment and hepatic function of transplanted cells is very hard which affects the success of the majority of the direct injection studies.

MICROCARRIER-ATTACHED HEPATOCYTE INJECTION

An alternative way to direct injection is the use of microcarrier-attached hepatocyte injection. Specifically engineered biodegradable microcarriers that contain extracellular matrix features can facilitate survival and function of hepatocytes, as well as their identification and analyses after transplantation. These microcarriers have very high surface area-to-volume ratio which permit a large number of cells to be transplanted. Type I collagen-coated dextran beads were evaluated as attachment surfaces for hepatocytes and encouraging results were obtained in the correction of metabolic defects in UDP-glucuronyl transferase deficient Gunn rats, and also in analbuminaemic Nagase rats, for over a month[11]. Subsequently, the survival time of 90% hepatectomized rats was significantly increased by using the same approach[22]. Intraperitoneal transplantation of microcarrier-attached hepatocytes has also been shown to significantly reduce the serum LDL levels in congenitally hyperlipidemic Watanabe rabbits[25]. In another study, fibrinogen-modified poly(hydroxyethylmethacrylate) microcarriers were developed for high hepatocyte immobilization. The hepatic function of attached cells was shown by urea and protein synthesis[26].

Biodegradable poly(lactide-co-glycolide) (PLGA) nanospheres that were surface modified by poloxamer and poloxamine copolymers were searched for their hepatocyte attachment ability and *in vivo* biodistribution in the rat and rabbit models after intravenous injection[27]. These hepatocyte carrying PLGA nanospheres exhibited prolonged blood circulation times accompanied by a combined reduction in liver and spleen accumulation after intravenous injection to the rat.

Among other transplantation sites, intraperitoneal engrafment showed the best results in microcarrier-attached hepatocyte injections, in terms of undetectable inflammation or toxicity and stimulation of cell organization and vascularization. However, immunosuppression was still needed to protect hepatocytes from body rejection[11,22].

MICROENCAPSULATED HEPATOCYTE INJECTION

Microcapsules are spherical, usually synthetic polymeric systems with semipermeable, ultrathin membranes. These biocompatible membranes may also contain extracellular matrix components that permit the identification and analysis of cell survival and function after transplantation. Microencapsulated hepatocyte systems have been used in several laboratory animals of hepatic dysfunction.

Alginate-poly(L-lysine) membranes have been used to microencapsulate hepatocytes[28]. By this way, microencapsulated cells could be shielded from immunological attack and a sustained reduction in the serum bilirubin was achieved in Gunn rats that received intraperitoneal injection, upto a month with increased survival times of rats subjected to galactosamine-induced fulminant hepatic failure[29].

In vivo studies with encapsulated hepatocytes showed encouraging results in the therapy of deficient liver functions without immunosuppression. The main shortcoming of these studies was the limited functional viability of the cells within a range of one to two months. A long term control of hyperbilirubinemia was achieved by using hepatocytes that were microencapsulated with collagen within alginate-poly(L-lysine) membranes[23]. By the repeated transplantation approach, the control of hyperbilirubinemia extended upto six months. Microcarrier-attached and microencapsulated hepatocyte systems seem to be effective at providing temporary hepatic support, however further research is needed for their long-term success.

BIODEGRADABLE SCAFFOLDS FOR HEPATOCYTE TRANSPLANTATION

The latest tissue engineering approach is the use of polymer scaffolds for the transplantation of hepatocytes. These biodegradable polymers have large surface area-to-volume ratios and can be designed to incorporate signals that affect hepatocyte regulation, function and reorganization. The highly porous surfaces also promote the attachment of a large number of cell mass which is crucial to provide hepatic support. Both the chemical composition of the surfaces, and the biomechanics of the cell-surface adhesion needs to be considered in designing such hepatocyte substrates. Ideally, the reorganization and expansion of transplanted cells should start during biomaterial degradation which may yield in the formation of the new liver organoid without chronic foreign body responses.

Poly(lactic acid) (PLA), poly(glycolic acid) (PGA), their blends, and their copolymers (PLGA) have been used as drug delivery devices[30,31]. This class of polymers have also been evaluated as substrates for hepatocyte growth and reorganization. Cell substrates made up of 50/50 (w/w) PGA and PLA blends were found to be suitable to extended hepatocyte culture and the cells retained differentiated function as measured by the rate of albumin secretion as well as DNA synthesis[32]. In another study, non-woven filamentous PGA sheets seeded with hepatocytes were transplanted to mesentery leaves of UDP-glucuronyl transferase deficient rats[33]. The devices showed moderate inflammatory reaction with high neovascularization and the transplanted cells were found to protect their specific function. The presence of bilirubin conjugates in Gunn rat bile profiles were also observed[33]. Poly(vinyl alcohol) (PVA) infiltrated PLA sponges with 90-95% porosity have been evaluated as three-dimensional hepatocyte scaffolds[34]. PVA infiltration was performed to enable even and efficient cell seeding by introducing this hydrophilic polymer into the sponges.

A recent study has described a new class of carbohydrate-modified hydrogels based on radiation-crosslinked star poly(ethylene oxide) (PEO)[35]. Hepatocytes exhibited a sugar-specific adhesion through the unique hepatic asialoglycoprotein receptors to galactose-modified gels with a considerable level of albumin production. In a similar approach, porous poly(styrene) foams were derivatized with lactose or heparin to promote hepatocyte attachment and function[36]. The cells on these substrates exhibited a higher level of albumin and cytochrome P_{450}-dependent hydroxytestosterone metabolite production compared to hepatocytes cultured on collagen surfaces for a week.

Acrylonitrile-sodium methallyl-sulfonate copolymer based hydrogel hollow fibers have also been used as hepatocyte transplantation devices[37]. Hepatocyte loaded hollow fibers were implanted into the peritoneum of Gunn rats without immunosuppression. A significant decrease in the serum bilirubin level, and bilirubin conjugates in bile samples were observed in subjects, with minimal fibrosis at the transplantation sites.

We have evaluated protein-modified forms of chitosan(*[1→4] linked 2-amino-2-deoxy-β-D-glucopyranose*) as specific attachment surfaces for rat and fetal porcine hepatocytes[9,38]. Protein modification was perfomed to enhance biocompatibility and surface roughness of the glycosaminoglycan analogue polymer[39,40]. In this study, while collagen-chitosan membranes provided the best attachment environment for the adult rat hepatocytes, albumin-chitosan membranes were found superior to fetal porcine hepatocyte attachment. The differences in attachment yields of the surfaces perhaps depend on the variable identificaiton of cell surface receptors that recognize and bind specific amino acid sequences present in extracellular matrix molecules[38,41]. Albumin-chitosan surfaces also provided a suitable environment for the metabolic activity of fetal porcine hepatocytes, in terms of urea and total protein secretion for at least two weeks. Chitosan-albumin scaffolds were used for the xenotransplantation of fetal porcine hepatocytes in subcutaneous tissue flaps of rats[42,43], after neovascularization of the transplantation area by a matrix-immobilized angiogenic growth factor, endothelial cell growth factor[44,45]. Previous studies primarily involved hepatocyte transplantation in sites such as spleen, directly into the liver via the portal vein, and the peritoneal cavity[8,13,17]. Using this technique, we explored the possibility of transplanting isolated fetal porcine hepatocytes to create liver tissue organoids at specific subcutaneous sites of rats without immunosuppression[42,43]. Immunohistochemical analysis also provided evidence for the creation of liver structures *de novo* from dissociated cells by using this polyelectrolyte-protein hydrogel system.

CONCLUSION

Surface chemistry and microstructure are important issues in tissue engineering studies that influence the ability of hepatocytes to attach, grow, and function as they do in their native tissues. The functional relevance of an *in vivo*-like extracellular matrix geometry for oxidative biotransformation of primary hepatocytes *in vitro*, may be defined in organotypical approaches[46]. The porosity and macroscopic dimensions of the cell substrates affect the transport of the nutrients to the implanted cells. The criteria of an implantation site should probably be its superficial location for practical processing, abundant vascularity for suitable nutritional support and the absence of other functional tissues that can interfere with the graft function. Also, the implant should contain a large number of hepatocytes that can give significant hepatic support. On the other hand, the number of seeded cells on three-dimensional scaffolds is also critical for successful transplantation, since the diffusional distance between the oxygen-carrying medium saturated with air and the cells should also be in a certain distance. Differential attachment to various substrates at intermediate seeding densities may be explained by specific interaction of the substrate with hepatocytes as a function of oxygen tension[30,47].

Mature hepatocytes may undergo of the order of fourteen doublings in an appropriate environment, however this number has not been reached with any tissue engineered liver organoid yet[31]. After isolation from intact liver, hepatocytes partially loose their phenotype and differentiated function[48]. Upon seeding onto or within a cell substrate, the cells undergo an adaptation and reorganization process. Another approach may be the use of fetal hepatocytes that have high proliferative properties[41,43,44]. It is assumed that fetal cells can adopt to the implantation environments and can be accepted by the host much easily than adult cells, since they do not have any HLA surface antigens[42,43]. An alternative approach may be the use of liver stem cells (oval cells) that are also known to be very replicative compared to mature hepatocytes[49]. These cells are more responsive to endogenous and exogenous hepatotrophic stimuli and may be considered as candidates for future liver tissue engineering studies[31].

An implantable tissue engineered liver organoid has not yet reached clinical realization. However, *in vitro* cultured hepatocytes on the surfaces or inside the scaffolds of biodegradable polymeric biomaterials still seem to be the most expectant way for successful transplantation.

Acknowledgments

The author expresses his thanks to Dr.s V. Dixit and G. Gitnick at U.C.L.A., Los Angeles, California, for their helpful suggestions and Turkish Scientific and Technical Research Council (TÜBİTAK) for the support of a part of this study.

REFERENCES

1. D.H. Van Thiel, L. Makowka, and T.E. Starzl, Liver transplantation: where it's been and where it's going, *Clin Gastroenterol.* 17:1 (1988).
2. P. Martin, S.J. Munoz, and L.S. Friedman, Liver transplantation for viral hepatitis: current status, *Am J Gastroenterol.* 87:409 (1992).
3. A.L. Jones, Anatomy of the normal liver, in: *Hepatology. A textbook of liver disease*, D. Zakim and T.D. Boyer, eds., W.B. Saunders Co., Philadelphia (1990).
4. B.M. McGuire, T.D. Sielaff, S.L. Nyberg, M.Y. Hu, F.B. Cerra, and J.R. Bloomer, Review of support systems used in the management of fulminant hepatic failure, *Dig Dis.* 13:379 (1995).
5. J. Gerlach, K. Klöppel, H.H. Schauwecker, R. Tauber, C. Müller, and E.S. Bücherl, Use of hepatocytes in adhesion and suspension cultures for liver support bioreactors, *Int J Artif Organs.* 12:788 (1989).
6. W.S. Arnaout, A.D. Moscioni, R.L. Barbour, and A.A. Demetriou, Development of bioartificial liver: bilirubin conjugation in Gunn rats, *J Surg Res.* 48:379 (1990).
7. V. Dixit, R. Darvasi, M. Arthur, K.J. Lewin, and G. Gitnick, Improved function of microencapsulated hepatocytes in a hybrid bioartificial liver support system, *Artif Organs.* 16:336 (1992).
8. V. Dixit, Development of a bioartificial liver using isolated hepatocytes, *Artif Organs.* 18:371 (1994).
9. Y.M. Elçin, V. Dixit, and G. Gitnick, Hepatocyte attachment on modified chitosan membranes, *Int J Artif Organs.* 18:464 (1995).
10. M.L. Yarmush, M. Toner, J.C.Y. Dunn, A. Rotem, A. Hubel, and R.G. Tompkins, Hepatic tissue engineering: development of critical technologies, in: *Biochemical Engineering VII*, H. Pedersen, R. Mutharasan, and D. DiBasio, eds., Ann. N.Y. Acad. Sci. Vol. 665, New York (1992).
11. A.A. Demetriou, J.F. Whiting, D. Feldman, et al., Replacement of liver function in rats by transplantation of microcarrier-attached hepatocytes, *Science.* 233: 1190 (1986).
12. S. Bruni, and T.M.S. Chang, Hepatocytes immobilized by microencapsulation in artificial cells: effects on hyperbilirubinemia in Gunn rats, *Biomater Artif Cells Artif Organs.* 17: 403 (1989).
13. P. Maganto, P.G. Traber, C. Rusnell, W.O. III. Dobbins, D. Keren, and J.J. Gumucio, Long-term maintenance of the adult pattern of liver-specific expression for P-450b, P-450e, albumin and α-fetoprotein genes in intrasplenically transplanted hepatocytes, *Hepatology.* 11: 585 (1990).
14. H.E. Rugstad, S.H. Robinson, C. Yannoni, and A.H.Jr. Tashjion, Transfer of bilirubin uridine diphosphate-glucuronyl-transferase to enzyme-deficient rats, *Science.* 170: 553 (1970).
15. A.B. Mukherjee, and J. Krasner, Induction of an enzyme in genetically deficient rats after grafting of normal liver, *Science.* 182: 68 (1973).
16. S. Saito, K. Sakagami, N. Koide et al., Transplantation of spheroidal aggregate cultured hepatocytes into the rat spleen, *Transplant Proc.* 21: 2374 (1989).
17. K. Onodera, H. Ebata, M. Sawa et al., Comparative effects of hepatocellular transplantation in the spleen, portal vein, or peritoneal cavity in congenitally ascorbic acid biosynthetic enzyme-deficient rats, *Transplant Proc.* 24: 3006 (1992).

18. A.J. Matas, D.E. Sutherland, M.W. Steffes, S.M. Mauer, A. Lowe, R.L. Simmons, and J.S. Najarian, Hepatocellular transplantation for metabolic deficiencies: decrease of plasma bilirubin in Gunn rats, *Science*. 192:892 (1976).

19. L. Makowka, R.E. Falk, L.E. Rotstein, J.A. Falk, N. Nossal, B. Langer, L.M. Blendis, and M.J. Phillips, Cellular transplantation in the treatment of experimental hepatic failure, *Science*. 210: 901 (1980).

20. P. Rivas, A.J. Fabrega, D. Schwartz, W. Digiantis, and R. Pollak, Preservation and transplantation of purified canine hepatocytes, *Transplant Proc*. 24: 2833 (1992).

21. M. Mito, and M. Kusano, Hepatocyte transplantation in man, *Cell Transplant*. 2: 65 (1993).

22. A.A. Demetriou, A. Reisner, J. Sanchez, S.M. Levenson, A.D. Moscioni, and J.R. Chowdury, Transplantation of microcarrier-attached hepatocytes into 90 % partical hepatectomized rats, *Hepatology*. 8: 1006 (1988).

23. V. Dixit, R. Darvasi, M. Arthur, K.J. Lewin, and G. Gitnick, Cryopreserved microencapsulated hepatocytes: transplantation studies in Gunn rats, *Transplantation*. 55: 616 (1993).

24. B.J. Fuller, Transplantation of isolated hepatocytes. A review of current ideas, *J Hepatol*. 7: 368 (1988).

25. J.C. Weiderkehr, G.T. Kondos, and R.Pollak, Hepatocyte transplantation for the low-density lipoprotein receptor-deficient state, *Transplantation*. 50: 466 (1990).

26. V. Dixit, E. Pişkin, M. Arthur, A. Denizli, S.A. Tuncel, E. Denkbaş, and G. Gitnick, Hepatocyte immobilization on PHEMA microcarriers and its biologically active forms, *Cell Transplant*. 1: 391 (1992).

27. S.E. Dunn, A.G.A. Coombes, M.C. Garnett, S.S. Davis, M.C. Davies, and L. Illum, *In vitro* cell interaction and *in vivo* biodistribution of poly(lactide-co-glycolide) nanospheres surface modified by poloxamer and poloxamine copolymers, *J Control Rel*. 44: 65 (1997).

28. V. Dixit, R. Darvasi, M. Arthur, M. Brezina, K. Lewin, and G. Gitnick, Restoration of liver function in Gunn rats without immunosuppression using transplanted microencapsulated hepatoyctes, *Hepatology*. 12: 1342 (1990).

29. H. Wong, and T.M.S. Chang, Bioartificial liver: implanted artificial cells microencapsulated living hepatocytes increases survival of liver failure rats, *Int J Artif Organs*. 9: 335 (1986).

30. Y.M. Elçin, Tissue engineering of liver, in: *Proc. Biomed-IV*, A. Hıncal, and S. Kaş, eds., İstanbul (1997).

31. M.W. Davis, and J.P. Vacanti, Toward development of an implantable tissue engineered liver, *Biomaterials*. 17: 365 (1996).

32. L.G. Cima, D.E. Ingber, J.P. Vacanti, and R. Langer, Hepatocyte culture on biodegradable polymeric substrates, *Biotechnol Bioeng*. 38: 145 (1991)

33. L.B. Johnson, J. Aiken, D. Mooney, B.L. Schloo, L.G.-Cima, R. Langer, and J.P. Vacanti, The mesentery as a laminated vascular bed for hepatocyte transplantation, *Cell Transplant*. 3: 273 (1994).

34. D.J. Mooney, S. Park, P.M. Kaufmann, K. Sano, K. McNamara, J.P. Vacanti, and R. Langer, Biodegradable sponges for hepatocyte transplantation, *J Biomed Mater Res*. 29: 959 (1995).

35. S.T. Lopina, G.Wu, E.W. Merrill, and L.G.-Cima, Hepatocyte culture on carbohydrate-modified star polyethylene oxide hydrogels, *Biomaterials*. 17: 559 (1996).

36. A.T. Gutsche, H.Lo, J. Zurlo, J. Yager, and K.W. Leong, Engineering of a sugar-derivatized porous network for hepatocyte culture, *Biomaterials*. 17: 387 (1996).

37. N. Gomez, P. Balladur, Y. Calmus, M. Baudrimont, J. Honiger, R. Delelo, A. Myara, E. Crema, F. Trivin, J. Capeau, and B. Nordlinger, Evidence for survial and metabolic activity of encapsulated xenogeneic hepatocytes transplanted without immunosuppression in Gunn rats, *Tranplantation*. 63: 1718 (1997).

38. Y.M. Elçin, V. Dixit, and G. Gitnick, Hepatocyte attachment on biodegradable modified chitosan membranes: *in vitro* evalution for the development of liver organoids, *Artif Organs*. submitted (1997).

39. A.E. Elçin, Y.M. Elçin, and G.D. Pappas, Neural tissue engineering: adrenal chromaffin cell attachment and viability on collagen-chitosan substrates, in: *Proc. Biomed-III*, Y. Ulcay, ed., Bursa (1996).

40. A.E. Elçin, Y.M. Elçin, and G.D. Pappas, Neural tissue engineering: survial and integration of bovine chromaffin cells on collagen-chitosan substrates transplanted into rats, in: *Proc. Biomed-IV*, A. Hıncal, and S. Kaş, eds., İstanbul (1997)

41. S.K. Akiyama, K. Nagata, and K.M. Yamada, Cell surface receptors for extracellular matrix components, *Biochim Biophys Acta*. 1031: 91 (1990).

42. Y.M. Elçin, V. Dixit, and G. Gitnick, Tissue engineering of bioartificial liver using biodegradable polymeric scaffolds: *in vitro* and *in vivo* studies, in: *Proc. Biomed-III*, Y. Ulcay, ed., Bursa (1996).

43. Y.M. Elçin, V. Dixit, K. Lewin, M. Arthur, and G. Gitnick, Tissue engineering studies involving xenotransplantation of fetal porcine hepatocytes in rats using biodegradable polymer scaffolds, *Cell Transplant.* submitted (1997).

44. Y.M. Elçin, V. Dixit, and G. Gitnick, Controlled release of endothelial cell growth factor from chitosan-albumin microspheres for localized angiogenesis, *Int J Artif Organs.* 18: 433 (1995).

45. Y. M. Elçin, V. Dixit, and G. Gitnick, Controlled release of endothelial cell growth factor from chitos an-albumin microspheres and fibers for localized angiogenesis: *in vitro* and *in vivo* studies, *Artif Cells Blood Subs Immob Biotechnol.* 24: 257 (1996).

46. A. Bader, E. Knop, N. Fruhauf, D. Crome, K. Boker, U. Christians, K. Oldhafer, B. Ringe, R. Pichlmayr, and K.F. Sewing, Reconstruction of liver tissue *in vitro*: geometry of characteristic flat bed, hollow fiber, and spouted bed bioreactors with reference to the *in vivo* liver, *Artif Organs.* 19: 941 (1995).

47. A. Rotem, M. Toner, S. Bhatia, B.D. Foy, R.G. Tompkins, and M.L. Yarmush, Oxygen is a factor determining *in vitro* tissue assembly: effects on attachment and spreading of hepatocytes, *Biotechnol Bioeng.* 43: 654 (1994).

48. J.A. Rhim, E.P. Sandgren, J.L. Degen, R.D. Palmiter, and R.L. Brinster, Replacement of diseased mouse liver by hepatic cell transplantation, *Science.* 263: 1149 (1994).

49. L. Germain, M. Noël, H. Gourdeau, and N. Marceau, Promotion of growth and differentiation of rat ductular oval cells in primary culture, *Cancer Res.* 48: 368 (1988).

AN INDEPTH CHARACTERIZATION OF BHK CELL LINES

Alison Stacey[1], Glyn Stacey[1], and Saime İ. Gürhan[2]

[1]European Collection of Cell Cultures (ECACC)
CAMR, Salisburry, Wilthshire SP 4 OJG, UK
[2]Animal Cell Culture Collection (HÜKÜK)
Foot and Mouth Diseases Institute
06044 Ankara, Turkey

INTRODUCTION

BHK cell lines are used for various purposes including virus detection and production of virus and recombinant proteins in many laboratories throughout the world[1]. They also are valuable tools in the biomaterials surface characteristics studies. However, clones of this cell-line perform variability of vital behaviours such as morphology, culture characteristics and susceptibility to some viruses.

Many BHK cell strains are available worldwide and as a generally accepted opinion the characterists of individual clones may vary considerably. In view of this we have undertaken an indepth study of a variety of characterists using 22 BHK clones from different sources worldwide. Parameters such as karyotype, growth cycle, population doubling time, transformation indices and suspectibilty to two virus species were studied besides isoenzyme profiles and DNA fingerprints.

MATERIALS AND METHODS

Cell Cultures

A panel of 22 BHK cell strains obtained from 4 laboratories were studied. GMEM (Gibco, UK with 10 % TPB, and 10 % FBS was used as growth medium. All cultures were subjected to quality control tests for the presence of mycoplasma, bacteria and fungi.

Growth Curves and Population Doubling Times (PDT)

Growth characterists of the cell strains were determined by both drawing the growth curves in identical incubation conditions. PDT were calculated by Equation 1.[2]

$$PDT = \frac{t}{n} = \frac{t \times 0.3}{\log Ct - \log Co} \qquad \text{(Equation 1)}$$

t = exponential growth period
n = generation number
Co = cell density at the beginnig of the exponential phase.
Ct = cell density at the end of the exponential phase

Karyotyping

Chromosome analysis of cells were carried out after metaphase arrest by incubation in the presence of 0.1 u g/ml colcemid (GIBCO, UK).

Isoenzyme Analysis

Preparation of Cell Lysate. Confluent cultures were harvested using 0.25 % trypsin / 10 mM EDTA sol. washed twice in ice cold PBS and resuspended in a little volume of buffer (10 mM EDTA, 100 mM Tris pH = 7, 2 % Triton X100) equivalent to size of cell pellet (50-100 ul). Following 10 minutes incubation on ice cell lysis assessed under a light microscobe. The lysed cells were centrifuged (400g, 4 °C, 10 min.) and clarified lysate was aliquoted, either used immediately or stored in vapour phase of liquid nitrogen.

Gel Preparation and Electrophesis. Analysis of cell lysates were carried out according to the manufacturers instructions (Innovative Chemistry, Marshfield, MA).[5] Lysates of HeLa S3 and L929 were included on every gel as a control and standart respectively.

DNA Fingerprinting. This was carried out according to the previously described method using the 33.15 DNA probe.[3,4] Briefly genomic DNA was extracted using phenol / chloroform and digested with Hinf1 (80 units / reaction in a total reaction volume of 42 ul). DNA from HeLa S3 and K562 was incubated in parallel as standarts and controls. The resutant DNA fragments were then separeted by electrophoresis at 50-60 mA (constant current) at room temparature for 18-20 h until a 2.4 kb marker (Marker IV, Boeringer Manheim, Germany) had migrated a distance of 20 cm. Gels were then Southern blotted and hybridised using the NICE[TM] system (Cellmark Diagnostics, UK).

Transformation Indices (% CFE). Colony forming efficiency of each cell strain was evaluated by cultivation in a double agar layer 4 % agarose, (Sigma, USA) as described previously[1] and calculated by Equation 2.

$$\% \ CFE = \frac{\text{total number of colonies}}{\text{total number of cell seeded}} \times 100 \qquad \text{(Equation 2)}$$

Virus Suspectibilty Studies

Foot and Mouth Disease Virus (FMDV) Suspectibilty Tests. Two different serotypes (A_{22} / Mah / 65 and O_1/ Man/69) of the virus was used for that purpose. The test was designed as described before.[2] Cultures sampled six hour intervals. Infectivity titres of the samples were carried out according to the cytopathogenic effects following inoculation of ten fold dilutions of viruses on swine kidney (IBRS$_2$) cells grown in 96 well tissue culture dishes.

Rabies Virus Suspectibilty Tests. In this test field strain of rabies virus (CVS- II, BHK , BSR P5) was used. The design of the test was similiar to FMDV test except the sampling intervals were 12 h and infectivity titres estimated by immunoflurescence microscopy according to the percentage flurescence positive cells.[6]

RESULTS AND DISCUSSION

Cell Growth and Morphology

The mean PDT of 22 BHK cell strains was 7.1 h ranging between 4.6-15.3 h (data for BHK An $_{52}$ excluded). Mean maximum cell density in 25 cm^2 culture flasks was 0.88 x 10^6 (range 0.11- 2.97 x 10^6). Cell morphologies varied from fibroblast -like to spherical as well as culture characteristis. 11 cell strains-typical fibroblast-like anchorage dependent (M), 6 cell stains-mixed morphology mostly fibroblast-like grew as both monolayer and suspension systems but need adaptation period for suspension (M and M+S), 2 cell strains equal number of fibroblast-like and spherical cells anchorage independent (S).

Species Identification Markers

Metaphase chromosome number for Syrian hamster is 2n = 44. Percetage of cells exhibiting this varied from 4 (BHK clone 13 IZS) to 52 BHK C13 An I [5], most lines displayed divers karyotypes. All clones were hypodiploid, few cells in each clone were seen to be hyperdiploid, none seen in BHKC13 An$_{39}$, C13-3P C13, AC9. Only 2 cells observed with karyotype 44, An$_{47}$ (equivalent to AC9) and An$_{39}$. Karyotype of clone 3P lines especially An$_{31}$ and An $_{32}$ is relatively stable with a minor differences between these 2 clones. These 2 clones were obtained from the same laboratory but with a two year acquisaions. Clone 13-3P obtained from same source but 3 years latter than An$_{31}$ deposit was obtained. All lines verified as Syrian hamster by isoenzyme analaysis. All cell clones had smiliar fingerprint patterns with only one or two band differences for any pair of cultures. All compared with ECACC line clone 13 (BHK21 C13).

Virus Suspectibilty

All clones were suspectible to FMDVs (A$_{22}$ and O$_1$) variable degrees. Cells were less suspectible to O$_1$ type than A$_{22}$. Only 8 clones were susceptable to rabies virus.

Colony Forming Efficiency

CFE of four clones were less than 10 %. Six clones were between 14 to 52 % and twelve clones were in correlation with morphology and culture charactertics.

Table 1. BHK cell-line strain characterization

| NAME BHK | GROWTH CURVE | | | MORPH. | TRANS (CFE %) | KARYO (2n=44) % | Isoenz | DNA FP | FMDV | | Rabies |
	MCD $x\ 10^4\ /\ 25\ cm^2$	P.D (h)	Stationary phase						A_{22}	O_1	CVS-II
An5	88.9	6.2	+	M	100	52	Sry. H.	BHK	++++	+++	+++
An10	76.7	6.2	+	M	50	-22	Sry. H.	BHK	+++	+++	+
An13	12.0	5.6	-	M + S	100	40	Sry. H.	BHK	++	++	++
An28	35.2	9.5	+	M + S	100	8	Sry. H.	BHK	+++	+++	+
An30	161.0	6.4	+ / -	M + S	52	14	Sry. H.	BHK	+++	++	+
An31	297.0	7.9	+	M	48	24	Sry. H.	BHK	+++	+	+
An32	11.4	15.3	+	M	50	2	Sry. H.	BHK	+++	+++	++
An39	76.0	6.4	+	M	14	44	Sry. H.	BHK	++++	++	+
An41	60.0	6.5	-	M + S	100	36	Sry. H.	BHK	+++	++	++
An47	113.3	5.7	-	M + S	100	44	Sry. H.	BHK	++	++	+++
An48	118.9	6.2	-	S	100	40	Sry. H.	BHK	++	+++	+
An52			+	M	4	24	Sry. H.	BHK	++	++	-
C13-3P	47.8	6.5	+	M	100	22	Sry. H.	BHK	++	++	++++
C13-2P	65.5	5.9	-	S	100	8	Sry. H.	BHK	++	+	+
TK -	97.7	13.0	+	M	4	8	Sry. H.	BHK	++	++	++++
C13	43.6	6.3	+	M	4	12	Sry. H.	BHK	++	+++	+
AC9	120.0	6.8	-	M + S	100	10	Sry. H.	BHK	+++	+++	-
CC2	77.8	6.2	+	M	14	16	Sry. H.	BHK	+++	+++	+
C13-IZS	94.5	6.8	-	M	7	4	Sry. H.	BHK	++	+++	+++
BS31	70.5	4.6	-	M + S	100	46	Sry. H.	BHK	+++	++	+
BS35	132.8	6.0	-	M + S	100	6	Sry. H.	BHK	+++	+++	+
BS38	144.4	5.4	-	S	100	32	Sry. H.	BHK	+++	+++	+

FMDV: $< 10^{-6.5}$ +
$10^{-6.6} - 10^{-7.0}$ ++
$10^{-7.1} - 10^{-7.5}$ +++
$> 10^{-7.6}$ ++++

Rabies : $< 30\%$ +
31-59 %   ++
60-79 %   +++
80-100 %   ++++

CONCLUSIONS

*PDT does not seem to vary with morphology of cells

*Karyotype is less stable in clones which grown in suspension or exhibit mixed population (M+S) (That is important for production of viral recombinant proteins)

*Variation in terms of growth and karyotype seen in clones cultured in laboratory and clones derived from same stocks but in different laboratories indicates that BHK clones are inherently unstable and not as result of scientific technique

*BHK clones are suspectible to FMDVs than rabies

*Virus susceptibilty is also variable between clones should be considered as an important factor in the planning of virus tests

REFERENCES

1. N. Bouch, and G. DiMayorca, 1979, Evaluation of chemical carcinogenicity by in vitro neoplastic transformation, in: *Methods of Enzymology,* Volume 58 (W.B. Jakoby, and I.H. Pastan, eds.), pp. 269-302.

2. S.İ. Gürhan, B. Gürhan, H. Öztürkmen, 1987, Examination of the growth chracterictics and suspectibilty of the cell cultures derived from BHK-21 populations by cloning to foot and mouth disease virus, *J. Doğa, TU Vet. Anim.* **11** (1): 21-27.

3. A.J. Jeffreys, V. Wilson, and S.L. Thein, 1985, Individual-spesific fingerprints of human DNA, *Nature* **316:** 76-79.

4. G.N. Stacey, 1991, in: *DNA Fingerprinting Approaches and Applications (T. Burke ed.),* Birkenhauser, Berlin.

5. G.N. Stacey, B.J. Bolton and A. Doyle, 1992, DNA fingerprinting transforms the art of cell outoidentification , *Nature* **8:** 13-20.

6. P. Stohr, K. Stoh, H. Kingel, E. Kargo, 1992, Immunofluorescence microscopy applied to field isolates of rabbies virus from various reagions of eastren Germany, using several fluorescein-labeled conjugates, *Tierarztliche Umschau* **47**(11): 813-818.

7. G.N. Stacey, B.J. Bolton, D. Morgan, S.A. Clark, A. Doyle, 1992, Multilocus DNA fingerprint analysis of cell banks: Stability studies and culture identification in human B-lymphoblastoid and mammalian cell lines, *Cytotecnology* **8:** 13-20.

INVESTIGATION OF BIOLOGICAL AND POLYMERIC MATERIAL USING ATOMIC FORCE MICROSCOPY

Hadi Zareie,[1] A. Patrick Gunning,[2] Ferdi Özer,[1] E. Volga Bulmuş,[1] Andrew R. Kirby,[2] A. Yousefi Rad,[1] A. Kevser Pişkin,[1] Vic J. Morris,[2] and Erhan Pişkin [1]

[1]Chemical Engineering Department
 Bioengineering Division
 Hacettepe University
 TÜBİTAK-Centre of Excellence:Polymeric Biomaterials
 Beytepe, Ankara, Turkey
[2]Institute of Food Research Norwich Laboratory
 Norwich Research Park
 Colney, Norwich NR4 7UA, UK

INTRODUCTION

The atomic force microscope (AFM) is a valuable imaging technique that is making inroads in various disciplines of biomedical researches. The applications of this new technologies to biomedical fields and materials science has been actively pursued by devoted researchers.[1-4] Unlike the scanning tunnelling microscope (STM), the AFM does not require the specimen to be either electron or ion conductive. Therefore, it has a much broader application, not only in conductive materials, but also in non conductive materials research. To date, the AFM at room temperature has been used to image various biological and polymeric material specimens, such as DNA,[5] membrane proteins,[6] cells,[7] polymers.[8] Potentially, polymeric and biological macromolecules are interesting candidates for study by atomic force microscope. This article presents images of biological and polymeric material by this technique. virus particles have been considered as a calibration specimen for SFM studies,[9] and here we used AFM to image the papaya mosaic virus (PMV) isolated from plant cells. In this image we have observed the leaking of RNA from virus. The immune response is mediated by antibodies, which act to bind a diverse range of foreign molecules. As described in detail by Roitt et al.,[10] higher mammals have five classes of immunoglobulin: IgA, IgD, IgE, IgG, IgM. Of the five distinct classes of immunoglobulins we chose to study IgG and IgM because these antibodies are used for other purposes in our laboratories, and the structure of these antibody have been the most thoroughly investigated. Atomic force microscopy images of these antibodies have been obtained in this study. Polymerization in microemulsions is a new polymerization technique which allows the synthesis of ultrafine latex particles within the size range 5nm < radius < 50 nm and with narrow size distribution.

In contrast to the opaque and milky conventional emulsions and miniemulsions, microemulsions are isotropic, optically transparent, and thermodynamically stable. They can be formed spontaneously by mixing oil and water in the presence of appropriate amounts of surfactant and cosurfactant (short chain alcohol or amine) and do not require vigorous agitation, high pressure homogenization, or ultrasonication.[11] Polystyrene particles formed by microemulsion polymerization technique have been imaged by atomic force microscopy in order to determine their size distribution.

MATERIALS AND METHODS

The viruses were a gift from Dr. Brian Wellss of John Innes Center, Norwich UK. IgG/IgM solution were purchased from Biotest (Germany). Styrene monomer (Sigma, USA) was purified by washing with 10% aqueous NaOH solution. The purified styrene monomer was stored at 4 °C until use. 1- pentanol (BDH, UK) and water-soluble initiator potassium persulfate (KPS) (BDH, UK) were used as supplied. The emulsifier used was a reagent-grade dodecylsulfate sodium (SDS) (Sigma, USA). Distilled water was used as polymerization medium. In a typical polymerization experiment, styrene monomer (6.22% based on total polymerization mixture) and 1-pentanol (3.95%) were mixed and the mixture was added to the 8.90 wt% aqueous solution of SDS in a 125 ml polymerization reactor. The contents of the reactor were agitated at room temperature for 10 min. Then it was stored at 4 °C for 24 hr under nitrogen atmosphere to attain equilibrium. After the addition of initiator (KPS, 0.02%) to the polymerization mixture, polymerization was carried out at 70 °C for 24 h under nitrogen atmosphere.

After polymerization, poly(styrene) latexes were separated from the polymerization mixture by precipitating in methanol and washed several times with methanol and water in order to remove physically adsorbed surfactant and cosurfactant. Size distributions of latex particles were determined by AFM.

The AFM used for the present studies was manufactured by ECS (East Coast Scientific, Cambridge, UK). The samples were contained in a liquid cell and imaged using constant force conditions. Optimum imaging conditions used force ~3-4 nN. The tips used were the short narrow variety of nanoprobe cantilevers (Digital Instruments, Santa Barbara, CA) with a nominal force constant of 0.38 Nm^{-1}. For all samples a 2µl drop were deposited onto freshly cleaved mica and allowed to dry in air for 10 min. The samples were then imaged under butanol (Sigma Chemical Co., St. Louis, MO). All quoted sizes are measured widths and overestimates due to probe-broadening effects.

RESULTS AND DISCUSSION

Representative AFM images of papaya musaic virus (PMV) are shown in Fig. 1A,B. In Fig. 1A, the virus particles appeared distributed in a monolayer over the mica surface; contaminations were rarely observed. The image has been taken in dc contact (Repulsive) mode under butanol, the scan area in Fig1A is 4 x 4µm. The papaya mosaic virus is known to be a long flexous rod with a length of 510 nm and a diameter of 11.5 nm. The variation in lengths seen in the images (Fig. 1A,B) probably arises due to break down of the virus partivles upon storage and clustering. Fig. 1B, shows another area of the same sample of PMV molecules on mica. This slightly high-resolution image shows both looping and leaking of RNA from virus particles. If the virus is at its isoelectric point, RNA looping out from the virus can occur.

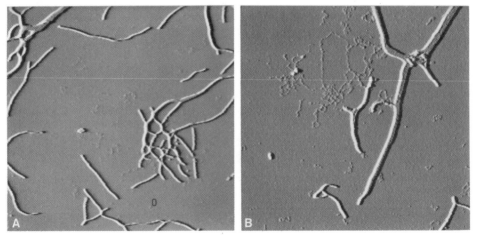

Figure 1. DC contact atomic force microscopy (AFM) error-signal mode images of papaya mosaic virus (PMV) on mica under butanol: (A) Scan size: 4 x 4 μm. (B) Scan size: 3 x 3 μm. Imaging force~3-4 nN.

Figure 2.(A,B) are representative AFM images of antibodies (IgG, IgM) on cleave mica and imaged under butanol. As is evident from figs. 2A and 2B, the method of deposition of antibodies provides a uniform distribution of macromolecules on the surface. Because the antibodies have a tendency to produce aggregates deposition from very dilute solution is useful in obtaining images of individual molecules. Although distorted by tip interactions, Fig. 2A suggests that the IgG molecules appear as non-spherical objects. The image size in Fig. 2A is 300 x 300 nm and shows a region of the surface that contains several IgG molecules. All the molecules were approximately of the same size and some could be taken to show a characteristic (Y) shape formed by the two Fab arms and the Fc region. Some of the IgG molecules may differ in shape because of differing orientation of the molecules on the mica surface. The shapes may be blurred by molecular motion or tip distortion. The size of the IgG molecules are measured approximately 35-50 nm. This increased size could be due to the presence of remaining solvent around the molecule, deformation by AFM tip during imaging or most likely due to probe-broadening. Fig. 2B, shows representative AFM images of IgM on mica under butanol. Immunoglobulin M (IgM) accounts for approximately 10% of the immunoglobulin pool, but are the major antibodies of the primary immune response. IgM antibodies corresponds roughly to a pentameric form of the basic IgG molecule together with a so-called J-chain. Fig.2B shows many IgM molecules deposited on the surface of mica, the scan area in this image is 800 x 800 nm. The size for on IgM molecule as measured from the AFM images, is approximately 60-80 nm: again this dimensions are larger than expected primarily due to probe-broadening.

In contrast to the opaque and milky conventional emulsions and miniemulsions, microemulsions are isotropic, optically transparent, and thermodynamically stable. Unlike the situation in standart emulsion polymerization, the droplet size in microemulsions is thermodynamically controlled by the amount and character of the surfactant. In this study, poly(styrene) spherical latexes paarticles approximately 20-40 nm in diameter were produced by microemulsion polymerization technique, using SDS as a surfactant. Since a large amount of surfactant is used during the microemulsion polymerization stage, purification of the resulting latex particles is an important step in the preparation prior to use of these particles. The AFM images of both the unpurified and purified (washed with

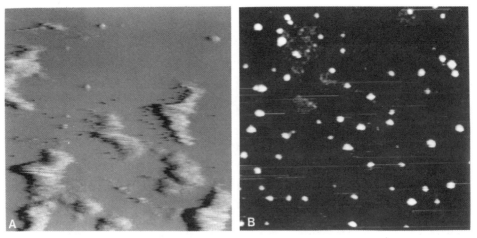

Figure 2. (A) AFM image of IgG molecules on mica under butanol: scan size is 300 x 300 nm. (B) AFM image of IgM molecules on mica: scan size is 800 x 800 nm.

methanol and water after polymerization) poly(styrene) latex particles have been obtained (Fig. 3). Fig. 3A shows representative image of pure SDS deposited from solution and imaged under butanol. This allows us to discriminate between SDS (micelles) and polystyrene latex particles. The scan size for Fig.3A is 300 x 300 nm. The widths of SDS particles are found to be 15 nm which is much smaller then the sizes expected for the latex particles. Fig.3B,C shows AFM images of unpurified and purified latex particles, respectively. For the unpurified sample (Fig. 3B) the majority of the surface is covered by SDS molecules and other components from the preparation process. The latex particles can also be seen in the image but represent a small percentage of the sample. The scan size for this image is 1.5 x 1.5 μm. The AFM image for purified microparticles is shown in Fig. 3C. The scan size for this image is 800 x 800 nm. The image shows that the mica surface is completely covered by latex particles. The aim of the study is to monitor the size of the latex particles, since characterization is important for their final application. The measured widths of the particles are between50-70 nm. For the close-packed arrays shown in Fig. 3C the probe-broadening effects will be minimized and these are realistic diameters for the particles.

CONCLUSIONS

As a conclusion we can strongly say that AFM is an analytical technique for the characterization and manipulation of biological molecules and polymeric surfaces. With AFM, one can image molecules, molecular complexes and touch them and act on them. This sense of actually touching a single molecule gives a totally new feeling to researchers, prompting them to explore the world of atoms and molecules with sense of innovation. Our understanding of biological molecules and polymer surfaces is much improved by performing experiments on the scanning force microscopy.

Acknowledgments

Dr. Hadi Zareie acknowledge support for the present research from the Underwood Fund, Biotechnology and Biological Sciences research Council (BBSRC) UK. Research on probe microscopy at IFR is funded by the BBSRC.

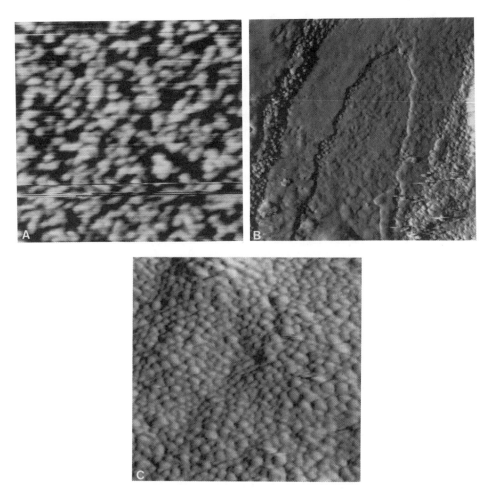

Figure 3. (A) AFM image of pure SDS deposited from solution and imaged under butanol: scan size 300 x 300 nm. (B) AFM error-signal mode image of unpurified polystyrene latex particles; scan size 1.5 x 1.5 μm. (C) AFM error-signal mode image of purified polystyrene latex particles; scan size of 800 x 800 nm.

REFERENCES

1. E. Engel, 1991, Biologication applications of scanning probe microscopes, *Annu. Rev. Biophys. Biophys. Chem.* **20**: 79-108.
2. H. G. Hansma, J. H. Hoh, 1994, Biomolecular imaging with the atomic force microscope, *Annu. Rev. Biophys. Biomol. Struct.* **23**: 115-139.
3. V. J. Morris, 1994, Biological application of scanning probe microscopies, *Prog. Biophys. Mol. Biol.* **61**: 131-185.
4. A. P. Gunning, A. R. Kirby, V. J. Morris, B. Wells, B. E. Brooker, 1995, Imaging bacterial polysaccharides with AFM, *Polym. Bull.* **34**: 615-619.
5. C. Bustamante, J. Vesenka, C. L. Tang, W. Rees, M. Guthod, 1992, Circular DNA molecules imaged in air by scanning force microscopy, *Biochemistry.* **31**: 22-26.
6. J. Yang, L. K. Tamm, T. W. Tillack and Z. Shao, 1993b, New approach for atomic force microscopy of membrane proteins: the imaging of cholera toxin, *J. Molec. Biol.* **299**: 286-290.

7. J. H. Hoh and P. K. Hansma, 1992, Atomic force microscopy for high resolution imaging in cell biology, *Trends Cell Biol.* **2**: 208-213.

8. R. M. Overney, 1995, Nanotribological studies on polymers, *TRIP.* **11**: 359-364.

9. T. Thundat, X-Y. Zheng, S. L. Sharp, D. P. Allison, R. J. Warmack, D. C. Joy and T. L. Ferrell, 1992, *Scanning Microscopy.* **6**: 903-910.

10. I. Roitt, J. Brostoff and D. Male, 1985, *Immunology*, pp. 5.1-5.10.

11. M. Antonietti, R. Basten, S. Lohman, 1995, Polymerization in microemulsions- A new approach to ultrafine, highly functionalized polymer dispersions, *Macromol. Chem. Phys.* **196**: 441-466.

THE FUTURE POTENTIAL FOR THE USE OF ADJUVANTS IN HUMAN VACCINES

Duncan Stewart-Tull

Division of Infection and Immunity
Institute of Biomedical and Life Sciences
Joseph Black Building
University of Glasgow
Glasgow, G12 8QQ, U.K.

INTRODUCTION

After some seventy years of research on adjuvants (immuno-stimulators, immuno-potentiators, immuno-modulators) it may seem difficult to comprehend why so few have been actively pursued for use in vaccines. There are numerous explanations but among those which seem to be of greatest concern to pharmaceutical companies are:
a) the unacceptability of substances isolated from pathogenic organisms,
b) the problems encountered in preparing large quantities of the pure, synthetic, active substance,
c) the cost of production of the adjuvant and hence the projected cost of the vaccine
d) the possible levels of inherent toxicity of a new vaccine formulation and the associated problem of litigation arising from the use of such a vaccine.

These criteria have caused the demise of a number of experimentally highly-active, adjuvant-active compounds. It is essential to measure the levels of reactivity and toxicity of the constituent antigens and immunostimulators to be used in experimental vaccines before proceeding to immunological studies. As reported recently at a workshop on nucleic acid vaccines " toxicology studies should be performed to evaluate local and systemic effects including inflammation and necrosis associated with components present in the final formulation such as facilitators like adjuvants or cationic lipid delivery systems (Smith et al 1997). At one international meeting it was stated by some speakers that "adjuvants are too reactive for inclusion in vaccines " (Stewart-Tull and Brown 1993), this is an extreme view, mostly propounded by those more concerned with regulatory issues than with laboratory experimental studies. It would seem more sensible to test for the adverse effects or contra-indications before expending more time on biological experimental vaccine work. The need to examine the toxicity levels by a number of alternative methods has been proposed (van Zutphen and Balls, 1997) and in my own work I aim to use the scheme in Figure 1.

Another hindrance has been the proposal for a single universal adjuvant, experience has revealed that within the collection of immunomodulators no one would prove to be beneficial in all vaccines. Many experimental vaccines contain different types of candidate antigen, from dead whole organisms to recombinant products of a pathogen. The use of these with a single immuno-stimulator has often resulted in poor stimulation of adjuvanted humoral or cell-mediated responses and adverse reports of the chosen immuno-potentiator appear in the literature. Greater stress must be placed on the necessity to test a variety of immuno-stimulators and delivery vehicles with a new vaccine antigen(s). In one adjuvant trial five different immuno-potentiators were tested with one antigen, the two most active were ruled out due to cost implications and the third most active was chosen for use in an

Biomedical Science and Technology
Edited by Hıncal and Kas, Plenum Press, New York, 1998

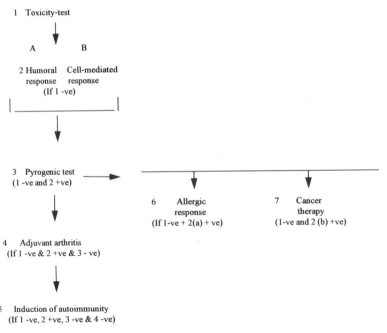

Figure 1. Sequence of testing immunopotentiating agents. If a particular substance fails to produce an acceptable result in the system described, there would be no further testing. Procedures 9-15 relate to specific requirements for the development of anti-bacterial vaccines. These procedures are carried out as part of the exercise to demonstrate vaccine safety or efficacy, where applicable.

experimental vaccine. Sometimes inter-company competition in the search for a new immuno-stimulator and the financial rewards a new licenced immuno-stimulator would bring may also lead to constraints in the search for good vaccine adjuvants. Notwithstanding all of these matters it is true that a great deal of experimental work has been completed, some one hundred adjuvants have been recorded but few immunostimulators have been licenced and designated for use in human vaccines. There is only time in this short presentation to mention a few which illustrate some of the difficulties and constraints which must be faced.

ALUMINUM ADSORBENTS

The exceptions are of course the aluminum salts, especially aluminum hydroxide and aluminum phosphate gels, although sometimes grouped together under the term 'alum', which have been used for more than fifty years in the form of alum-precipitated or alum-adsorbed vaccines for human and animal use. Since the introduction of the diphtheria-pertussis-tetanus vaccine millions of doses have been injected into babies and young children with no small amount of success. There is still a place for the use of aluminum salt if the conditions are such that the antigen has the correct charge to permit adsorption to produce the 'alum-precipitated vaccines' (Stewart-Tull, 1996a). Seeber et al. (1991) demonstrated a difference in solubility between aluminum hydroxide and aluminum phosphate, the latter dissolved more rapidly which indicated that the hydroxide was probably more effective in maintaining the depot for the slow release of antigen. These salts also differ in their protein adsorptive capacity, that for aluminum hydroxide was 1.8 - 2.6 mg ml^{-1} Al, whereas aluminum phosphate was 1.0-1.3 mg ml^{-1} (Hem and White 1995). It is possible to make the alum salts in the laboratory but I prefer to use a standardised, commercial preparation produced by Superfos Biosector a/s in Denmark in my experimental vaccines. .The main drawback against the use of aluminum salts is that such vaccines containing adsorbed antigen may be useful in stimulating a humoral immune response, with the stimulation of IL-4 and Th2 subsets with increased production of IgG1 and IgE antibodies, but they are not very active in stimulating cell-mediated responses. For example, sometimes authors will claim that aluminum salts are inferior to their own immunopotentiator, while ignoring the fact that they do not stimulate a cell-mediated immunity or they report that alum salts are implicated in Alzheimer's disease. The latter is very misleading especially as genetical studies have shown that the disease is gene related; In all probability many people will have been exposed to much higher levels of aluminum from eating green cabbage cooked in aluminum pots with some soda in the water to maintain the green colour.

Where a strong Th2 response is required then aluminum salts have proved to be very effective in preventing diseases caused by soluble, protein toxins which can be neutralised by anti-toxin antibodies. It follows however, that an aluminium salt would not be the immunomodulator of choice where a Th1 response was required to effect a cure, for example in tuberculosis or leishmaniasis. Major advantages are the lower cost of production and hence the economic value per dose of vaccine, ~300 tons of aluminum hydroxide gel are used annually, and the suitability for large-scale vaccine production.

IMMUNE - STIMULATING OIL EMULSIONS

The classical Freund's complete adjuvant (FCA), composed of heat-killed cells of *Mycobacterium tuberculosis* in a mineral oil phase and the protein antigen in an aqueous phase mixed with an emulsifier to form a stable water-in-mineral oil emulsion, or the Freund's incomplete adjuvant, formulated without the mycobacterial component, have received a bad press during the past few years. In some countries the use of FCA is now positively discouraged, even for experimental use, because of adverse tissue reactions. These were due in part to the use of crude mineral oils which contained a mixture of short- and long-chain hydrocarbons obtained as by-products of the petroleum industry and in part to the unacceptable use of a pathogen, albeit a dead organism, known to stimulate granulomatous lesions.

Some recent studies indicate that there may still be a future for 'Freund-type incomplete adjuvant mixtures'. The hydrocarbon mineral oils used prior to 1970 were produced by the acid treatment or oleum method. These crude oils contained paraffins (alkanes, saturated non-cyclic hydrocarbon chains) cycloparaffins (naphthenes, saturated cyclic hydrocarbon rings) olefins (unsaturated hyrdocarbon chains) and aromatic hydrocarbons. For instance, Drakeol 6-VR was a mineral oil composed of a complex mixture of normal, cyclic and branched-chain hydrocarbons and aromatics. Monobutyl and dibutylphthalate were present in small amounts, (Murray et al., 1972). Since then they have been produced by the single or double hydrogenation procedure and superior grades of oil have been obtained. These new oils were far less reactive and less toxic, Stewart-Tull (1995) but these have also been superceded by some non-toxic oils which are biodegradable. The time has come to stop designating all oil emulsions with the name of Jules Freund to avoid the tremendous confusion which arises by association with the specific formulations of Freund studied in the 1930s and 1940s. These new formulations should be divorced from the old terminological link and designated as '**Immune-stimulating oil emulsions**' (IOEs). Stewart-Tull (1996b) described new formulations, reportedly without unacceptable levels of toxicity or local injection site adverse reactions, which could be used in experimental systems. One of these oils, namely Montanide ISA 720 (Seppic, Paris), was used in a Phase 1 trial in Australia and the side-effects were minimal and short-lived; immediately after injection into the thigh or deltoid muscle some of the volunteers experienced 'slight tenderness or discomfort' for 1 to 4 days There were no apparent ill-effects during the one-year follow-up of the twelve volunteers who received between 0.6 and 1.8ml doses by deep intra-muscular injection into the thigh. They concluded that this trial was encouraging as some of the volunteers received three times the dose which would be used in a vaccine in one injection (Lawrence et al., 1996). Montanide ISA 720 contains a highly-refined emulsifier in a natural metabolizable oil used as 70 parts Montanide ISA 720 : 30 parts antigen in an aqueous phase. After emulsification the mixture is stable for many months on the shelf but it is essential to make up the emulsion in the proportions stated above otherwise it is very unstable. In a study with influenza

A/Beijing/353/89 (H3N2) sub-unit vaccine, Benne et al., (1997) showed that Montanide ISA 740 predominantly stimulated an IgG1 response after six weeks with lower titres of IgG2a and IgG2b in female BALB/c mice, however, FCA (Difco) and L180.5 (non-ionic block co-polymer) / RaLPS (Sigma) appeared to stimulate all three isotypes. and Q-VAC (a mixture of saponins) stimulated mainly IgG1 and IgG2a. This reveals the importance of examining different antigens with different immuno-potentiators.

Another oil-in-water emulsion which is undergoing field trials is MF 59, this is composed of squalene (a triterpene from shark liver oil or vegetable oil which is an intermediate in cholesterol biosynthesis; 43.0 mg ml^{-1}), polysorbate 80 (polyoxyethylene sorbitan monooleate; 2.5mg ml^{-1}) and Span 85 (sorbitan trioleate; 2.4mg ml^{-1}) and has been recommended for intra-muscular injection. Squalene may be harmful by inhalation or ingestion, the oral LD$_{50}$ is 5g kg^{-1}; therefore the toxicity at the level of a vaccine dose may require investigation before use in an oral or nasal vaccine to stimulate mucosal immunity. It is also worth noting other differences of MF 59 with 'Freund-type emulsions'; with FCA vegetable oils had low adjuvant activity (Woodhour et al., 1961) and in addition water-in-oil emulsions were usually more active than oil-in-water emulsions. The difference with MF 59 may be due to the use of purer materials or to the microfluidization at 12000 psi which yielded droplets with a mean diameter of <300nm which were stable in physiological saline. Some experimental vaccines also used squalane, which is the total hydrogenation product of squalene, this was not as active alone but was used in combination with other active components, see SAF-1. From experimental studies with MF 59 the titres of antibody were > than those stimulated by aluminum hydroxide, = or > than those from FIA stimulated, and = or < than those from FCA primary with a booster of FIA (Ott et al,1995). In general, polysaccharides by themselves are not helped by adjuvants, however, a recent study used a conjugate vaccine in baboons which contained *Neisseria meningitidis* group C and *Haemophilus influenzae* type b oligosaccahride-CRM$_{197}$ conjugates in aluminum hydroxide or MF 59. Serologically cross reacting proteins were produced by chemical mutagenesis tachniques in the 1970s. They are non-toxic but still react with the native toxin and induce protective immunity. CRM197 was obtained from *Corynebacterium diphtheriae*, it is serologically identical to native exotoxin but was naturally toxoided due to the induction of a missense mutation in the A fragment gene by treatment with nitrosoguanidine (Pappenheimer, 1984) The animals were given an injection of 0.5ml containing ~10µg of each saccharide and a total of 40µg of the CRM$_{197}$ protein, subsequently at monthly intervals, they were given two booster injections of 0.5ml containing 5.0µg of the unconjugated *N.meningitidis* group C polysaccharide vaccine (Chiron) and 5.0µg of Hib polysaccharide vaccine (Chiron) in PBS. The MF 59 injected animals had up to seven-fold higher geometric mean titres of capsular antibodies than the alum inoculated group (Granoff et al., 1997).

NATURAL AND SYNTHETIC PEPTIDOGLYCAN OR LIPOPOLYSACCHARIDE DERIVATIVES

It must be said that the era of research into the immunopotentiating activity of the native and synthetic polymers associated with the cell-wall of microorganisms has not yielded a wealth of immunopotentiators suitable for use in vaccines. Billions of BCG doses have been administered since the vaccine was introduced and each dose would contain approximately 3.0-5.0 mg of peptidoglycan. Whole cell, Gram-negative bacterial vaccines contain both peptidoglycan and lipopolysaccharide which are known adjuvants, (Stewart-Tull, 1985). Both synthetic muramyl dipeptide derivatives (Ellouz et al., 1974) and Lipid A (Van Boeckel et al., 1982) have immuno-potentiating activity . However, many of the muramyl dipeptide derivatives possessed high levels of toxicity and were only used as laboratory tools. At present MTP-PE, which has reduced levels of toxicity, is being tested in MF59 (4.3% w/v squalene, 0.5% w/v Tween 80, 0.5% w/v Span 85 (a mixture of Arlacel 85 and sorbitan trioleate) in a microfluidised emulsion with droplets of ~400 nm in diameter (Ott et al, 1995) . This formulation was safe and stimulated both humoral and cell-mediated responses with subunit or recombinant antigens from herpes simplex, influenza and human immunodeficiency viruses and is being tested in Phase I-III human field trials.

SAPONINS AND IMMUNE-STIMULATING COMPLEXES

In regard to veterinary vaccines, especially viral vaccines (Rimmelzwaan and Osterhaus, 1995) a first line approach would be to test immune-stimulating complexes (ISCOMs) which contain saponin components from the South American soaptree *Quillaja saponaria* Molina, designated as Quil A (Dalsgaard, 1978; Dalsgaard et al., 1995) but many derivatives are to be found in other plants (Campbell ,1995). Kensil et al., (1991) purified the crude mixture of saponins and obtained a fraction referred to as QS-21 which had similar adjuvant properties. Neuzil et al., (1997) examined a number of immuno-potentiators, namely 10μg QS-21, 10μg monophosphoryl lipid A (MPL) and 100μg alum or combinations of these, especially FG/MPL/QS-21 formulated in liposomes, in a detailed study with 10μg respiratory syncytial virus FG subunit vaccine. BALB/c mice were injected intramuscularly with 200μl of the vaccines and challenged after 30 days with live virus. The composites were more effective than the single immunopotentiators in aiding the recovery of the animals. In addition, FG/MPL/alum stimulated the highest titre of IgG after 56 days, the lowest number of plaque-forming units per gram of lung and the highest titres of both IgG1 and IgG2a when compared to experimental vaccines containing QS-21.Nevertheless, only the intra-nasal instillation of the live virus resulted in the production of virus neutralizing antibodies. These authors produced results in this experimental mouse model which indicated the beneficial and additive effects of combining adjuvants. Although the work of Benne et al., (1997) mentioned above indicated that Q-VAC, was less active with influenza antigens, with blood-stage antigens of *Plasmodium yoelii* elevated levels of IgG1,IgG2a,IgG2b and IgG3 were stimulated against the particulate antigen (pAG). Protection against infection, due to IgG1 and IgG2b, was comparable when pAG was used with Quil A or with RAS (50μg of monophosphoryl lipid A plus 50μg of synthetic trehalose dicorynomycolate: RIBI) or with FCA/FIA (Burns et al., 1997).

The ISCOM forms as a honeycomb, cage-like sphere, 30-40nm in diameter, when the Quil A, cholesterol and phospholipids together with the antigen are mixed under specific conditions (Dalsgaard et al., 1995; Stewart-Tull,1996c) Some researchers found this difficult to achieve but subsequent work showed that it was possible to preform the empty spheres and then add the antigen. Whether or not the known haemolytic effect of saponins will exclude the use of Quil A and its derivatives, either alone or in the form of ISCOMs, in human vaccines remains to be seen.

LIPOSOMES AND MICROPARTICLE DELIVERY SYSTEMS

Another delivery system which has attracted attention is the incorporation of antigens into artificial lipid bilayers, liposomes. These can be used to administer an antigen in conjunction with an immunopotentiator associated with the lipid bilayer. The liposome has amphiphilic properties with both hydrophobic (phospholipids) and hydrophilic components. Initially, there were problems in making stable, unilamellar spheres, this could be explained in part by the impurities present in many commercial preparations of phospholipid. In experimental studies in my own laboratory it was routine to examine the constituents by thin layer chromatography for single spot phospholipid preparations. Invariably, impurities were found but this was overcome by the preparation of pure lecithin ourselves by the sophisticated method of 1955. Nevertheless, the experimental evidence revealed considerable instability in the artificial lipid bilayers (Stewart-Tull et al., 1973a,b) This problem has again been reiterated by Kourilsky (1996) to quote

" Phospholipids depending on the degree of unsaturation of the fatty acid tails, tend to oxidize with time and therefore require careful monitoring during storage. Though antioxidants can be used, this may present regulatory hurdles. .It has also been reported that under harsh conditions lyso-phospholipids can be generated. Experimentally, it has been shown that if the lysophospholipid and the free fatty acid generated are below ~1% of the total lipid, these two components continue to remain within the liposomal bilayer. However, above this percentage the lysophospholipid tends to leave the bilayer and can perturb other liposomal membranes. Small liposomes

($\leq$ 50nm) tend to fuse with other small liposomes in order to attain a more stable size. This is a critical problem for vaccine formulation. If the antigen is anchored in the membrane, there tends to be very little antigen loss. Fusion can also be triggered by external agents, even by the vaccine antigen itself. This is of particular concern when viral glycoprotein surface antigens used. Furthermore, liposomal fusion can also lead to change in lamellarity. If the antigen is anchored in the membrane, then the addition of an extra bilayer can lead to its masking and thereby affect its accessibility to the immune system."

Because of these difficulties attempts to make experimental vaccines required the use of a two vial system. The pure lipids were prepared as a dehydrated film in one vial and the antigen in aqueous solution in another. The vaccine was prepared just prior to inoculation by resuspending the lipid film with the antigen solution, such a system was deemed to be uneconomical and difficult to maintain good quality control when used for mass inoculation. Since these early studies the technologies have advanced and it is now possible to produce more stable liposomes. Consequently, there is renewed interest in the use of liposomal vaccines (Gregoriadis, 1996) It is encouraging to note that a liposome-based (virosome) inactivated hepatitis A vaccine has undergone phase I-III clinical trials and is now licenced for use in Switzerland.

The work on liposomes was the forerunner of development studies on the use of particulate vehicles for the presentation of antigens in microspheres and microparticles (O'Hagan, 1994; Cleland, 1995; Hanes et al., 1995; Kreuter, 1995; Payne et al., 1995; Santiago et al., 1995) This topic has been addressed by others in papers presented at this symposium and the reader is referred to other sections of this book.

CYTOKINES AND TOXINS

There is an extensive literature on the effects of cytokines and their possible use as immuno-potentiators. Cytokines, if produced by a lymphocyte referred to as lymphokines, are proteins which are produced by many cells and which play an important role in immune responses by either stimulating or suppressing cells involved in the immune system At present, I am cautious about their general use in immune reactions because our knowledge of cytokines tends to be supplemented each year. In 1979, some 39 immunologists signed a letter to the *J.Immunology* proposing that many cytokines should be grouped into IL-1 and IL-2, but since then there has been a continual increase in the list and many of the newer ILs have been shown to be involved in adjuvanted responses. The interactions which occur after the injection of ILs may have considerable effects on other cytokines and cellular processes, some of which are still to be discovered, and upset the steady state systems in and between immune cells. This does not exclude their potential use as immunotherapeutic agents in situations where there may be an insufficiency (Sadlack et al., 1993; Kahn et al., 1993). Greater knowledge about how these cytokines are involved in disease processes must be accumulated (Casey et.al., 1993) before their use as vaccine additives can be accurately assessed.

Similarly, with carrier proteins in conjugated vaccines it is important to consider the immune response to the carrier. *Vibrio cholerae* toxin (CT) and *Escherichia coli* heat-labile toxin (LT) or their detoxified forms are used in experimental vaccines but they cause diarrhoea at low concentrations when given orally to human beings (Levine et al., 1983). They are also good antigens and there is a query about how many doses could be given to an individual in different vaccines. When conjugated to a protein antigen antibodies would probably be produced against the toxin or its subunit, although some believe that this would not arise due to low dose tolerance.

CONCLUSIONS

The European Commission for Co-operation, Science and Technology in Europe and Life Sciences and Technologies for Developing Countries (1996) concluded that it was necessary to determine the optimal procedures by which antigens, including DNA, are processed and presented to immune cells in combination with immuno-stimulatory substances during immuno-potentiation of the immune response required to combat an infection.. In this respect it will be important to harness the requisite specific response ie Th1 (IFN-g and IL-2), Th2 (IL-4 and IL-10) or CTL with immuno-potentiators and delivery systems administered by the best route ie, parenteral, oral, nasal, rectal or vaginal or a combination of these preferably with a limited number of doses. There is a great potential in this area for research with a number of different strategies as no single substance has so far been found which meets all requirements. Some of the strategies used by the workers in the studies mentioned here may provide an incentive to explore the range of immuno-potentiators which exist, provided that the single or combined formulations do not induce toxic reactions.

REFERENCES

Benne, C.A., Harmsen, M., van der Graaff, W., Verheul,A.F.M., Snippe H., and Kraaijeveld, C.A. 1997. Influenza virus neutralizing antibodies and IgG isotype profiles after immunization of mice with influenza A subunit vaccine using various adjuvants. *Vaccine*.15:1039-1044

Burns, J.M., Dunn, P.D., and Russo, D.M. 1997. Protective immunity against Plasmodium yoelii malaria induced by immunization with particulate blood-stage antigens. *Infect and Immun.* 65: 3138-3145

Campbell, J.B. 1995 Saponins. in. *The Theory and Practical Application of Adjuvants,* D. E. S. Stewart-Tull, ed., John Wiley, Chichester, pp 95-127.

Casey, L.C., Balk, R.A., and Bone, R.C. 1993. Plasma cytokines and endotoxin levels correlate with survival in patients with sepsis syndrome. *Ann.Intern.Med.* 21:771-778.

Cleland, J.L. 1995 Design and production of single-immunization vaccines using polylactide polyglycolide microsphere systems. In. *Vaccine Design:The Subunit and Adjuvant Approach,* M.F.Powell and M.J. Newman, eds., Plenum Press, New York, pp 439-462.

Dalsgaard, K. 1974 Saponin adjuvants III. Isolation of a substance from *Quillaja saponaria*Molina with adjuvant activity in foot-and-mouth disease vaccines. *Aich. Ges.Virus.Forschung.* 44:243-254.

Ellouz, F.A., Adam, R., Ciobaru, R., and Lederer, E. 1974 Minimal structural requirements for adjuvant activity of bacterial peptidoglycan derivatives *Biochem. Biophys. Res. Commun.*. 59:1317-1325.

Granoff, D.M., McHugh, Y.E., Raff, H.V., Mokatrin, A.S., and Van Nest, G.A. 1997. MF 59 adjuvant enhances antibody responses of infant baboons immunized with *Haemophilus influenzae* type b and *Neisseria meningitidis* group C oligosaccharide-CRM197 conjugate vaccine. *Infect. and Immun.*65: 1710-1715

Gregoriadis, G 1996 Liposomes as immunological adjuvants in. *The Theory and Practical Application of Adjuvants,* D.E.S.Stewart-Tull, ed., John Wiley, Chichester, pp 144-169.

Hanes, J.,Chiba, M. and Langer, R. 1995 Polymer microspheres for vaccine delivery. in *Vaccine Design:The Subunit and Adjuvant Approach,* M.F.Powell and M.J. Newman, eds., Plenum Press,New York, pp 389-412.

Hem, S.L., and White, L. 1995 Structure and properties of aluminum-containing adjuvants. in *Vaccine Design:The Subunit and Adjuvant Approach,* M.F.Powell and M.J. Newman, eds., Plenum Press,New York, pp.249-276.

Kahn, R., Lohler, J., Rennick D., et al 1993. Interleukin-10-deficient mice develop chronic enterocolitis. *Cell.* 75:263-274.

Kensil, C.R., Patel, U., Lennick, M., and Marciani, D. 1991 Separation and characterization of saponins with adjuvant activity from *Quillaja saponaria* Molina cortex. *J.Immunol.* 146: 431-437.

Kourilsky, P. 1996. Stability of vaccine formulations for vaccine delivery. *Vaccine*.14: 688-689.

Kreuter, J. 1995 Nanoparticles as adjuvants for vaccines. in. *Vaccine Design:The Subunit and Adjuvant Approach,* M.F.Powell and M.J. Newman, eds., Plenum Press, New York, pp 463-472..

Lawrence, G. W., Saul, A., Giddy, A. J., Kemp, R., and Pye, D. 1997 Phase I trial in humans of an oil-based adjuvant SEPPIC MONTANIDE ISA 720. *Vaccine.* 15: 176-178.

Levine, M.M., Kaper, J.B., Black, R.E., and Clements, M.L. 1983. New knowledge on pathogenesis of bacterial enteric infections as applied to vaccine development. *Microbiol.Rev.*47:510-550

Murray, R., Cohen, P., and Hardegree, M.C. 1972 Mineral oil adjuvants: biological and chemical studies. *Anns of Allergy,* 30: 146-151.

Neuzil, K.M., Johnson,J.E., Tang, Y-W., Prieels, J-P., Slaoui, M., Gar, N., Graham, B.S. 1997 Adjuvants influence the quantitative and qualitative immune response in BALB/c mice immunized with respiratory syncytial virus FG subunit vaccine. *Vaccine.* 15: 525-532.

O'Hagan, D. T. 1994 Microparticles as oral vaccines. in *Novel delivery systems for oral vaccines.* D.T.O'Hagan, ed., CRC Press, New York, pp175-205.

Ott,G., Barchfield, G.L., Chernoff, D., Radhakrishnan, R., van Hoogevest, P., and Van Nest, G. 1995 MF59:Design and evaluation of a safe and potent adjuvant for human vaccines. in *Vaccine Design:The Subunit and Adjuvant Approach,* M.F.Powell and M.J. Newman, eds., Plenum Press,New York, pp 277- 296.

Payne, L.G., Jenkins, S. A., Andrianov, A., and Roberts, B. E. 1995 Water-soluble phosphazene polymers for parenteral and mucosal vaccine. in *Vaccine Design:The Subunit and Adjuvant Approach,* M.F.Powell and M.J. Newman, eds., Plenum Press, New York, pp 473-493.

Rimmelzwaan, G. F. and Osterhaus, A.D.M.E 1995 A novel generation of viral vaccines based on the ISCOM matrix. in. *Vaccine Design:The Subunit and Adjuvant Approach,* M.F.Powell and M.J. Newman, eds Plenum Press,New York, pp 543-558.

Sadlack, B., Merz, H., Schorle, H., et al 1993 Ulcerative colitis-like disease in mice with a disrupted interleukin-2 gene. *Cell.* 75: 253-261.

Santiago, N., Haas, S. and Baughman, R. A. 1995 Vehicles for oral immunisation. in *Vaccine Design:The Subunit and Adjuvant Approach,* M.F.Powell and M.J. Newman, eds., Plenum Press,New York, pp 413-438.

Seeber, S. J, White, J. L., and Hem, S. L. 1991 Solubilization of aluminum-containing adjuvants by constituents of interstitial fluid. *J.Parenteral Sci.Tech .* 45: 156-159

Smith, H.A., Goldenthal, K.L., Vogel F.R., Rabinovich, R., and Aguado, T. 1997 Workshop on the control and standardization of nucleic acid vaccines. Vaccine. 15:931-933.

Stewart-Tull, D. E. S., and Davies, M. 1985 Immunopotentiating activity of peptidoglycan and surface polymers in. *Immunology of the Bacterial Cell Envelope,* D.E.S.Stewart-Tull and M.Davies, eds., John Wiley, Chichester, pp. 47-89

Stewart-Tull, D. E. S., and Brown , F. 1993 First Steps Towards an International Harmonization of Veterinary Biologicals , *Vaccine .*11: 692-695.

Stewart-Tull, D. E. S. 1995 Freund-type mineral oil adjuvant emulsions. in. *The Theory and Practical Application of Adjuvants,* D.E.S.Stewart-Tull, ed., John Wiley, Chichester, pp 1-19.

Stewart-Tull, D. E. S. 1996a The use of adjuvants in experimental vaccines. I. Aluminum hydroxide gels. in. *Methods in Molecular Medicine: Vaccine Protocols* A. Robinson, G.H.Farrar and C.N.Wiblin,eds., Humana Press, New Jersey, pp.135-140.

Stewart-Tull, D. E. S. 1996b The use of adjuvants in experimental vaccines: II. Water-in-oil emulsions: Freund's complete and incomplete adjuvants, in, *Methods in Molecular Medicine: Vaccine Protocols .* A. Robinson, G.H.Farrar and C.N.Wiblin, eds., Humana Press, New Jersey, pp141-146.

Stewart-Tull, D. E. S. 1996cThe use of adjuvants in experimental vaccines. IV. ISCOMS. in.*Methods in Molecular Medicine: Vaccine Protocols.* A. Robinson, G.H.Farrar and C.N.Wiblin, eds., Humana Press, New Jersey, pp153-155.

Van Boeckel, C. A. A., Hermans, J. P. G., Westerduin, P., Oltvoort, J. J., van der Marel, G.A., and van Boom, J. H. 1982 Chemical synthesis of diphosphorylated lipid A derivatives. *Tetrahedron Lett..* 23, 1951-1954.

van Zutphen, L. F.M., and Balls, M. 1997. Eds *Animal Alternatives, Welfare and Ethics.* Elsevier, Holland.

Woodhour, A.F., Jensen, K.E., and Warren, J. 1961.Development and application of new parenteral adjuvants. V. Comparative potencies of influenza vaccines emulsified in various oils. J.Immunol. 86, 681-689.

EVALUATION OF THE ANTIBODY RESPONSE AFTER ORAL IMMUNIZATION BY MICROPARTICLES CONTAINING AN ANTIGEN FROM SCHISTOSOMA MANSONI

Marie-Ange Benoit,[1] Benoît Baras,[1] Odile Poulain-Godefroy,[2] Anne-Marie   Schacht,[2] André Capron,[2] Jean Gillard,[1] and Gilles Riveau[2]

[1] Université Catholique de Louvain
   Ecole de Pharmacie, Laboratoire de Pharmacie Galénique
   Industrielle & Officinale - Av. E. Mounier
   73.20 - 1200 Bruxelles, Belgium
[2] Institut Pasteur de Lille
   Centre d'Immunologie et de Biologie Parasitaire
   INSERM U 167 - Rue du Prof. Calmette
   1 - 59019 Lille, France

INTRODUCTION

The use of vaccines, broadly defined as agents that confer protective immunity against a pathogen, has had a significant effect on both human and animal health for more than two centuries with the Jenner's experiments and more recently with the Pasteur's discoveries. Nevertheless, vaccination remains more a deliberate art of empirical immunization than a real science in many instances, and scientific strategies of modern vaccinology have been developed only in the last few years. Even if a number of new approaches to vaccine development have emerged like recombinant subunit vaccines, synthetic epitopes and DNA vaccines, a general problem remains : the limited immunogenicity of these antigens. Consequently, many of the new-technology vaccines will also require adjuvants (which enhance the immune response to a vaccine) or delivery systems (which modify the temporal or spatial distribution of a vaccine) to allow them to induce protective immunity successfully.

Within this framework, the humoral adjuvant effect (characterized by the specific production of antibodies) achieved by entrapment of antigens in biodegradable microparticles has been demonstrated only relatively recently[1,2]. Moreover recent studies in mice have shown that microparticles also enhance the cell-mediated immunity by induction of a cytotoxic T lymphocyte (CTL) response against an antigen from HIV-1 after systemic and mucosal administration [3,4]. A delayed-type hypersensitivity (DTH) response and potent T-cell proliferative responses also occur when mice are treated with ovalbumin entrapped into microparticles[3].

Poly(lactide) (PLA) and poly(lactide-co-glycolide) (PLG) are the primary polymeric candidates for the development of microparticles as vaccine delivery systems because they are biocompatible and biodegradable polymers. Moreover, PLG polymers have been used in humans for many years as suture material and as controlled-release delivery system for peptide drugs[5]. In the long term, one of the most attractive features of these biodegradable carriers is their ability to control the rate of release of entrapped antigens. Ultimately, this could allow the development of single-dose vaccines through the preparation of microparticles that release entrapped antigen at the time points when booster doses of vaccine would normally be administered. So, a single dose of microparticles with entrapped protein from *Schistosoma mansoni* has been shown to induce a specific humoral response for at least 4 months[6]. Nevertheless, one of major drawback of these polymers is that their degradation generates extreme acid environment (pH 2-3) in which many antigens are found to lose their structural integrity and antigenicity[7,8].

Poly(ε-caprolactone) (PCL) is an other biocompatible and biodegradable polyester polymer that degrades slowly and does not generate an acid environment unlike the PLA/PLG polymers. Others advantages of PCL include its hydrophobicity, its in vitro stability and its low cost. Moreover, clinical trials of PCL-based contraceptive delivery system capronor™ (levonorgestrel) worldwide have demonstrated the safety and efficacy of this polymer[9].

The administration of vaccines via a mucosal route (oral, intranasal, intrarectal and inhalation) offers several significant advantages over the traditional approach to vaccine delivery. These advantages include easier administration, reduced side effects, the potential for frequent boosting without the need of trained personnel and the induction of mucosal immunity at the site of initial infection. Finally, since the protective barrier of the skin is not breached during administration, the potential for the introduction of infection through the use of contaminated needles is eliminated.

In mice, oral immunization with PLG microparticles induces potent serum IgG and secretory IgA antibodies, and CTL responses[3,10,11]. Although large doses of antigen were often used in these studies, a single oral dose of 10 μg of fimbriae from *Bordetella pertussis* in microparticles protected mice from intranasal challenge with the bacterium[12]. Moreover, in primates, intratracheal or oral delivery of microparticles with entrapped whole simian immunodeficiency virus (SIV) induced protective immunity against intravaginal challenge with the virus[13]. In a separate study in primates, intratracheal immunization induced protection against aerosol challenge with staphylococcal enterotoxin B[14]. Microparticles are therefore currently being evaluated in several Phase I clinical trials as oral antigen-delivery systems.

The purpose of this study was to encapsulate an antigen from *Schistosoma mansoni* in microparticles from two types of biodegradable polymers (PLG and PCL) by using a multiple emulsion solvent evaporation technique[15] for oral administration. This parasite is responsible for human schistosomiasis which is the second parasitic disease in the world after malaria causing 500,000 deaths per year in 80 countries. With a prevalence of 200 million people infected, schistosomiasis remains a major health problem in the developing world. Significant progress has been made towards the identification of protective antigen[16], and considerable attention has been directed to molecules common to the larval and adult stage of the parasite, such as the 28kDa protein characterized by an enzymatic activity related to glutathione S-transferase (Sm28GST) and that was found to protect rodents and baboons against experimental infections[17,18].

MATERIAL AND METHODS

Material

The following chemicals were obtained from commercial suppliers and used as received : Poly(ε-caprolactone) Tone E767E (80,000 M.W.) from Union Carbide (Versoix, Switzerland); Poly(lactide-co-glycolide) resomer RG 505 (33,000 M.W.) from Boehringer (Ingelheim, Germany); Polyvinyl alcohol (PVA) (13-23K, 87-89% hydrolyzed) from Aldrich Chemical Co (Bornem, Belgium); Dichloromethane (DCM) from UCB (Braine l'Alleud, Belgium); Sodium dodecyl sulfate (SDS), CDNB (1-chloro-2,4 dinitrobenzene) and GSH (glutathione) from Sigma (Bornem, Belgium); Horseradish peroxidase (HRP)-conjugated anti-mouse IgG isotypes from Southern Biotechnology Associates Inc.(Birmingham, Al, USA); Biotinylated goat anti-mouse IgA antibodies and streptavidin-horseradish peroxidase from Amersham (Les Ulis, France); ABTS (2,2'-azino-di-[3-ethyl-benzothiozolin-sulfonate]) from Boerhinger (Mannheim, Germany); Recombinant Sm28GST expressed in *Saccharomyces cerevisae* from Transgene (Strasbourg, France);

Tween, gelatin and Folin-Ciocalteu's phenol reagent and other materials (reagent grade) from Merck (Darmstadt, Germany).

Microparticle Preparation

An aqueous solution (1 ml) of Sm28GST was emulsified with a solution (10 ml) of PCL (6%, w/v) or PLG (5%, w/v) in dichloromethane using an ultraturrax model T25 (IKA Laboratory Technology, Staufen, Germany) at high speed (8,000 rpm) during 5 min. The resulting water-in-oil (w/o) emulsion (2.5 ml) was then emulsified (8,000 rpm for 5 min.) with a solution (50 ml) of PVA (5%, w/v) to produce a water-in-oil-in-water (w/o/w) emulsion. The w/o/w emulsion was then stirred magnetically overnight at room temperature and pressure to allow the evaporation of the organic solvent and the formation of microparticles. The microparticles were isolated by centrifugation (10 min. at 4,000xg), washed three times in 10 ml of ultra-pure water and freeze-dried.

Microparticle Characterization

The size distribution analysis of preparations was performed by using a Coulter Counter Multisizer (Coulter Electronics Ltd, Luton, UK) equipped with a sieve of 100 μm aperture and under continuous stirring. Particle size was expressed as volume mean diameter (VMD) in micrometers ± SEM (n=3).

The antigen content was determined by dissolving a sample in 3.0 ml of 1M NAOH containing 5% (w/v) SDS during 24h at room temperature. After centrifugation (4,000xg for 10 min), the supernatant was assayed following the method of Lowry[19]. The percentage (w/w) of antigen entrapped per dry weight of microparticles was determined. The percentage of entrapment efficiency was expressed by relating the actual antigen entrapment to the theoretical antigen entrapment.

Immunization

Six weeks old female BALB/c mice (Iffa Credo, L'arbresle, France) were intragastrically immunized by a single dose of the two types of microparticles as delivery systems or, in control groups, using a saturated solution of bicarbonate containing the same amount of free Sm28GST or empty microparticles. Each dose contained 100 μg of antigen in 200 μl of solution.

Blood samples and gut lavage fluids of mice were collected and treated as previously described[20].

Elisa

Purified Sm28GST (10 μg/ml in Phosphate Buffer Saline (PBS) at pH 7.4, 50 μl/well) was coated during 2h30 min at 37°C on 96 well Immulon 3 microtiter plates (Dynatech Laboratory Inc., VA, USA). The plates were washed three times with 0.1% (v/v) Tween 20 in PBS (PBS-Tween) using an automatic washer (LP35, Diagnosyic Pasteur, Marne-la-Coquette, France). Sera (including non-immunized mouse serum as control) diluted in PBS-Tween containing 0.5% (w/w) gelatin, were added to the antigen-coated wells (50 μl/well for each dilution) and incubated overnight at 4°C. Plates underwent four PBS-Tween washes and were incubated for 1h30 min. at 37°C with 50 μl of horseradish peroxidase (HRP)-conjugated goat anti-mouse IgG1 (1/1,000 dilution), IgG2a (1/500), IgG2b (1/500), IgG3 (1/300) appropriately diluted in PBS-Tween containing 0.5% (w/v) gelatin. After four washes, HRP was revealed for 1h at room temperature with 1 mg/ml ABTS in a citrate buffer (50 mM, pH 4.0) containing 0.003% H_2O_2. The absorbances were measured at 405 nm using a microplate reader (Titertek Multiscan MCC, Labsystems Group, Les Ulis, France).

For IgA determination in sera, ELISA protocol included one more step (after the incubation of sera) with biotinylated goat anti-mouse IgA antibodies diluted 1/3,000 in PBS-Tween containing 0.5% (w/v) gelatin for 1h30 min. at 37°C, washed six times, and incubated with streptavidin-horseradish peroxidase diluted 1/2,000 in PBS-Tween containing 0.5% (w/v) gelatin for 30 min. at 37°C. The plates were washed six times with

PBS-Tween, and 50 μl of peroxidase substrates (ABTS and H_2O_2) were added as described above.

Finally, for the measurement of IgGA in the gut lavage fluids, a saturation step with 100 μl of PBS containing 0.5% (w/v) gelatin for 30 min. at room temperature occurred after the coating of the antigen following by a single wash of plates. The samples were appropriately diluted in PBS-Tween and placed into the wells (50 μl) after three washes.

For each assay, the titers were defined as the highest dilutions yielding an absorbance three times above the background.

Neutralization of the Sm28GST Activity by Anti-Sm28GST Sera

The glutathione S-transferases are enzymes catalyzing conjugation reactions in which glutathione (GSH) acts as a nucleophile. The addition of GSH to the substrate bearing an electrophilic carbon atom (e.g. 1-chloro-2,4 dinitrobenzene, CNDB) results in the thioether formation and thus in a direct change in the absorbance of the substrate which is measured by spectrophotometry[21].

The neutralizing activity of an anti-Sm28GST serum was analyzed in 96 well flat-bottomed standard ELISA Plates as described by Grzych et al.[22], with the following modifications. In a total reaction volume of 430 μl containing 400 μl of reaction buffer, 30 μl of enzymatic mixture was added. This mixture contained 10 μl of recombinant Sm28GST (10 μg/ml) incubated with 20 μl of 50 mM potassium phosphate buffer (pH 6.5), or the studied antisera at different dilutions, for 1 h at 37°C followed by 1 h at 4°C. The reaction was started by the addition of the reaction buffer which was composed by 50 mM potassium phosphate buffer (pH 6.5) containing 0.36 mM CDNB and 4.76 mM GSH. The absorbances was recorded at 340 nm at 15 sec. intervals for a period of 2 min.

The enzymatic activity value recorded for the Sm28GST in presence of antisera was related to that measured with sera of non-immunized mice, which was assigned as the 100% value for specific activity. The results were obtained with pools of sera from five mice withdrawn 10 weeks after mucosal immunization.

RESULTS

Microparticle Characterization and Antigen Entrapment

Whatever the polymer used, the microparticles were spherical, smooth and fairly monodispersed. The main characteristics of the two types of microparticles are summarized in Table 1.

Table 1. Characteristics of microparticles from two types of polyester polymers entrapping Sm28GST

Polymer	Particle Size (μm)			Entrapment Efficiency (%)	Antigen Loading (%, w/w)
	Mean	Min.	Max.		
PCL	11.81 ± 0.31	6.01 ± 0.15	21.96 ± 1.93	20.0 ± 2.41	0.85 ± 0.18
PLG	9.94 ± 0.71	4.45 ± 0.10	16.07 ± 1.43	90.40 ± 2.84	3.33 ± 0.17

Each sample was assayed in triplicate.

The mean size of microparticles produced from the two polyester polymers was around 10 μm with an entrapment efficiency of $20.00 \pm 2.41\%$ and $90.40 \pm 2.84\%$ for the PCL and PLG samples with an antigen loading of $0.85 \pm 0.18\%$ and $3.33 \pm 0.17\%$ respectively.

Systemic Immune Response After Oral Administration

Oral immunizations with polymers entrapping recombinant Sm28GST, free antigen or empty microparticles were conducted by a single intragastric administration.

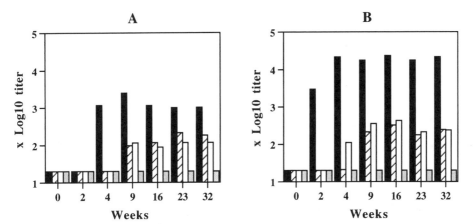

Figure 1. Antibody isotype profiles elicited after intragastric immunization with Sm28GST-microparticle complexes. Anti-Sm28GST IgG1 (black bars), IgG2a (hatched bars), IgG2b (white bars) and IgG3 (stipple bars) titers were determined at the indicated time points in pooled sera from ten BALB/c mice per group, before immunization (week 0) and after immunization (from week 2). Mice were immunized with either (A) PCL and (B) RG505 microparticles containing the antigen. Titers are given in Log10 of maximal dilution of antisera that gave an absorbance threefold higher than the background.

The serum IgG antibody responses to Sm28GST entrapped into PCL and PLG RG505 (Figure 1A & B) were highly detectable compared to responses obtained with empty microparticles or unencapsulated antigen that were totally ineffective (titer < 20, data not shown). In the two cases, the isotypic profile was characterized by the coexistence of IgG1, IgG2a and IgG2b without production of IgG3 or IgA. The predominant IgG1 immune response appeared 4 weeks after the oral immunization with PCL (Figure 1A) whereas specific IgG1 production started 2 weeks after PLG RG505 administration (Figure 1B). Both immune responses reached a plateau up to 24,000 (RG505-microparticles) and 2,500 (PCL-micropaticles) after 9 weeks. These responses did not decrease throughout the study which lasted 32 weeks. No IgG2a or IgG2b isotypes appeared before 4 weeks after the beginning of the IgG1 immune response.

Mucosal Immune Response After Oral Administration

IgA antibody responses in serum and gut lavage fluids resulting of the oral administration of PCL and RG505-microparticles loaded with Sm28GST or free antigen or empty microparticles were measured. Unlike the specific mucosal response induced by nasal administration[23] of these two carriers, no specific immune response characterized by the production of IgA into the sera and the gut lavage fluids was detected. The same results were also obtained when unencapsulated or empty microparticles were administered orally.

Antiserum-Mediated Neutralization of Sm28GST Enzymatic Activity

Anti-Sm28GST sera (week 10) obtained after a single oral immunization of mice with PCL and RG505-microparticles entrapping the antigen induced respectively a neutralization of 8.40 % and 45.1% of the enzymatic activity of the recombinant Sm28GST (Table 2). In contrast, no significant neutralization was detected with sera obtained after oral administration of free antigen or empty microparticles (< 5%).

Table 2. Neutralization of the Sm28GST enzymatic activity by an antiserum obtained after mucosal immunization of mice

Antigen formulation	Inhibition of the enzymatic activity (%)
Non-Immunized mice	0.00
Free Sm28GST	4.12
Empty microparticles	3.09
PCL Microparticles with Sm28GST	8.40
PLG Microparticles with Sm28GST	45.05

Results were obtained with pools of sera from five mice withdrawn 10 weeks after oral administration

DISCUSSION

A heat-stable, single-dose, non-toxic vaccine that could be given mucosally to protect effectively against schistosomiasis would be an important step towards increasing immunization coverage worldwide. This work was the first approach describing the potentialities of microparticles to elicit long-term immune responses after a single oral administration with a recombinant antigen from a multicellular and extracellular parasite.

Whithin the framework of previous studies, the potentialities of poly(ε-caprolactone) (PCL) and copolymers of lactic acid and glycolic acid (PLG) to encapsulate by a multiple solvent emulsion evaporation technique the Sm28GST from *Schistosoma mansoni* as well as the conservation of the structural integrity, antigenicity and immunogenicity of this entrapped antigen were already demonstrated[15,24].

Oral immunizations with Sm28GST entrapped within PCL- and RG505-microparticles elicited high and long-lasting humoral immune responses. Specific IgG1 antibodies were mainly induced with the coexistence of IgG2a and IgG2b isotypes. These results could suggest the induction of a mixed Th1-Th2 type response. In order to confirm the involvement of the two Th-cell subpopulations, the cytokine secretion profiles of antigen-stimulated CD4+ Th cells would be investigated. On the contrary, the oral administration of free-antigen was totally ineffective to enhance a specific immune response. This observation illustrated the ability of microparticles produced with these two types of polyester polymers to protect the loaded protein against the acidic and proteolytic environment of the gastro-intestinal tract.

The use of PCL microparticles allowed to delay the production of specific antibodies when compared to the response induced by PLG microparticles. Indeed, the predominant IgG1 isotype response elicited with the first type of microparticles appeared 2 weeks later than the one resulting of the administration of the PLG carriers. Morever, PCL might be absorbed most readily by the M Cells of the Peyer's patches due to its greater relative hydrophobicity. Futhermore, the low antigen entrapment within PCL microparticles when compared with PLG microparticles resulted in a higher number of PCL carriers administered to obtain the same final dose of 100 μg of Sm28GST. Thus, the delayed immune response observed with the first type of particles could be explained by 1) the high antigen retention in PCL microparticles correlated to its low permeability to macromolecules[9]; 2) the higher molecular weight of this polymer; 3) the difference of cristallinity between the two polymers or/and finally by the low antigen loading in PCL microparticles. These characteristics resulted in a slow antigen release rate and, consequently, an immune response which appeared later than for the PLG microparticles. So, the simultaneous administration of the two types of microparticles would be a first approach to produce a vaccine containing the first dose for the primary immunization (with PLG carriers) and the booster dose (with PCL carriers) following a single-step administration.

In this study, the oral administration of Sm28GST entrapped into microparticles was totally ineffective to enhance a high and specific mucosal responses whatever the polymer used. On the contrary, the same carriers were shown to produce IgGA into the blood and the broncho-alveolar lavage fluids after nasal administration[23]. This findings seem to be in accordance with other reports using multiple doses of various encapsulated antigens[25,26,27] where oral or systemic immunizations were required to prime the mucosal IgA response stimulated by booster administration via the oral route. Moreover, only few studies related that a single oral dose was sufficient to enhance mucosal immune responses[12,28] in rabbits or NIH-porton mice but no IgA response into the mucosal fluids was demonstrated until now in BALB/C mice. Finally, we believe that the different results could be correlated to the protocol of administration, the nature of the antigen and, principally, by the animal model used for the studies. Our results are also probably correlated to the particular features of GALT which is physiologically and anatomically different to the lymphoid tissue associated to the broncho-alveolar tract (BALT). By oral route, the microparticles which are not taken up by the GALT are eliminated in feces. On the contrary, microparticles administered by nasal route accumulate into the lung until their future uptake by the BALT. In this respect, the BALT would be several times re-stimulate even after a single nasal administration. Thus, the single oral administration of antigen-containing microparticles seems to be insufficient to allow the switch of B cells in IgA producing plasma cells.

Contrary to the results observed with the anti-Sm28GST antibodies obtained after administration of PCL microparticles, the antisera produced after the single oral immunization of mice with PLG RG505 were capable of neutralizing the enzymatic activity associated to Sm28GST. This result could be probably related to the difference of presentation or folding of the antigen into the two types of microparticles. Nevertheless, these results are of great interest because the Pasteur Institut of Lille (France) has been able to associate to the neutralization of the enzymatic activity by specific antibodies, a protection against schistosomiasis in humans[29] correlate to a dramatic reduction in egg laying and egg viability.

Acknowedgements

This work is supported by a grant N° BIO4-CT96-0374 from EEC. B. Baras is recipient of a FRIA (Fonds pour la Formation à la Recherche dans l'Industrie et l'Agriculture) fellowship. The size analyses of microparticles were kindly carried out at the Lab. Galenische en Klinische Farmacie (Prof. R. Kinget K.U.Leuven, Belgium).

REFERENCES

1. D.T. O'Hagan, D. Rahman, J.P. McGee, H. Jeffery, M.C. Davies, P. Williams, S.S. Davis and S.J. Challacombe, Biodegradable microparticles as controlled release antigen delivery systems, *Immunology* **73**: 239-242 (1991).
2. J.H. Eldridge, J.K. Staas, J.A. Meulbroek, T.R. Tice and R.M. Gilley, Biodegradable and biocompatible poly(DL-lactide-co-glycolide) microspheres as an adjuvant for staphylococcal enterotoxin B toxoid which enhances the level of toxin-neutralizing antibodies, *Infect. Immun.* **59**: 2978-2986 (1991).
3. K.J. Maloy, A.M. Donachie, D.T. O'Hagan and A.M. Mowat, Induction of mucosal and systemic immune responses by immunization with ovalbumin entrapped in poly(lactide-co-glycolide) microparticles, *Immunology* **81**: 661-667 (1994).
4. A. Moore, P. McGuirk, S. Adams, W.C. Jones, J.P. McGee, D.T. O'Hagan and K.H.G. Mills, Immunization with a soluble recombinant HIV protein entrapped in biodegradable microparticles induces HIV-specific CD8+ cytotoxic T lymphocytes and CD4+ Th1 cells, *Vaccine* **13**: 1741-1749 (1995).
5. D.L. Wise, T.D. Fellman, J.E. Sanderson and R.L. Wentworth, Lactide:glycolide polymers used as surgical suture material, raw material for osteosynthesis and in sustained release of drug, in: *Drug Carriers in Medicine*, G. Gregoriadis, ed., Academic Press, pp. 237-270 (1979).
6. B. Baras, M.-A. Benoit, B.B.C. Youan, G. Riveau, J. Gillard and A. Capron, Vaccine against schistosomiasis with spray-dried microparticles, *Proceed. Intern. Symp. Control. Rel. Bioact. Mater.* **24**: 815-816 (1997).
7. B. Gander, C. Thomasin, H.P. Merkle, Y. Men and G. Corradin, Pulsed tetanus toxoid release from PLA-microspheres and its relevance for immunogenicity in mice, *Proceed. Intern. Symp. Control. Rel. Bioact. Mater.* **20**: 65-66 (1993).
8. S.P. Schwendeman, H.R. Costantino, R.K. Gupta, M. Tobio, A.C. Chang, M.J. Alonso, G.R. Siberand and R. Langer, Strategies for stabilising tetanus toxoid towards the development of a single-dose tetanus vaccine, in: *New Approaches to Stabilisation of Vaccines Potency. Dev Biol Stand Basel*, F. Brown, ed., Karger, pp. 293-306 (1996).
9. C.G. Pitt, Poly(ε-caprolactone) and its copolymers, in: *Biodegradable Polymers as Drug Delivery Systems*, M. Chasin and R. Langer, eds., Marcel Dekker, pp. 71-120 (1990).

10. J.H. Eldridge, C.J. Hammond, J.A. Meulbroek, J.K. Stass, R.M. Gilley and T.R. Tice, Controlled vaccine release in the gut-associated lymphoid tissues. I. Orally administered biodegradable microspheres target the Peyer's patches, *J. Controlled Rel.* **11**:205-214 (1990).
11. S.J. Challacombe, D. Rahman, H. Jeffery, S.S. Davis and D.T. O'Hagan, Enhanced secretory IgA and systemic IgG antibody responses after oral immunization with biodegradable microparticles containing antigen, *Immunology* **76**:164-168 (1992).
12 D.H. Jones, B.W. McBride, C. Thornton, D.T. O'Hagan, A. Robinson and G.H. Farrar, Orally administered microencapsulated *Bordetella pertussis* protect mice from *B. Pertussis* respiratory infection, *Infect. Immun.* **64**:489-494 (1996).
13. P.A. Marx, R.W. Compans, A. Gettie, J.K. Staas, R.M. Gilley, M.J. Mulligan, G.V. Yamshchikov, D. Chen and J.H. Eldridge, Protection against vaginal SIV transmission with microencapsulated vaccine, *Science* **260**:1323-1327.
14. J. Tseng, J.L. Komisar, R.N. Trout, R.E. Hunt, J. Yok-Jen Chen, A.J. Johnson, L. Pitt and D.L. Ruble, Humoral immunity to aerosolized staphylococcal enterotoxin B (SEB), a superantigen, in monkeys vaccinated with SEB toxoid-containing microspheres, *Infect. Immun.* **63**:2880-2885 (1995).
15. M.-A. Benoit, B.B.C. Youan and J. Gillard, Potential of polyester microparticles for the sustained release of oral vaccine, *Proceed. 1st World Meeting on Pharmaceutics, Biopharmaceutics, Pharmaceutical Technology (APGI)* Budapest, 431-432 (1995).
16. D.W. Dunne, P. Hagan and F.G.C. Abath, Prospects for immunological control of schistosomiasis, *Lancet* **345**:1488-1491 (1995).
17. I. Wolowczuk, C. Auriault, H. Gras-Masse, C. Vendeville, J.-M. Balloul, A. Tartar and A. Capron, Protective immunity in mice vaccinated with the *Schistosoma mansoni* P28-1 antigen, *J. Immunol.* **142**:1342-1350 (1989).
18. D. Grezel, M. Capron, J.-M. Grzych, J. Fontaine, J.-P. Lecocq and A. Capron, Protective immunity induced in rat schistosomiasis by a single dose of the Sm28GST recombinant antigen: effector mechanisms involving IgGE and IgGA antibodies, *Eur. J. Immunol.* **23**:454-460 (1993).
19. O.H. Lowry, N.J. Rosebrough, A.L. Farr and R.J. Randall, Protein measurement with Folin phenol reagent, *J. Biol. Chem.* **193**:265-275 (1951).
20. N. Mielcarek, J. Cornette, A.-M. Schacht, R. Pierce, C. Locht, A. Capron and G. Riveau, Intranasal priming with recombinant *Bordetella pertussis* for the induction of a systemic immune response against a heterologous antigen, *Infect. Immun.* **65**:544-550 (1997).
21. W.H. Habig and W.B. Jakoby, Assays for differentiation of glutathione S-transferase, *Methods Enzymol.* **77**:398-405 (1981).
22. J.-M. Grzych, E. Pearce, A. Cheever, Z.A. Caulada, P. Caspar, S. Heiny, F. Lewis and A. Sher, Egg deposition is the major stimulus for the production of Th2 cytokines in murine schistosomiasis mansoni., *J. Immunol.* **146**:1322-1327 (1991).
23. M.-A. Benoit, B. Baras, L. Dupré, O. Poulain-Godefroy, A.-M. Schacht, A. Capron, J. Gillard and G. Riveau, Nasal vaccination against schistosomiasis using biodegradable microparticles, *Proceed. Intern. Biomed. Sci. Technol. Symp.* Istanbul, 27-28 (1997).
24. M.-A. Benoit, O. Poulain-Godefroy, B. Baras, B.B.C. Youan, G. Riveau, J. Gillard and A. Capron, Study on the antigenicity of microencapsulated Sm28GST from *Schistosoma mansoni, Proceed. Intern. Symp. Control. Rel. Bioact. Mater.* **24**:817-818 (1997).
25. J.H. Eldridge, J.K. Staas, J.A. Meulbroek, J.R. McGhee, T.R. Tice and R.M. Gilley, Biodegradable microspheres as a vaccine delivery system, *Mol. Immun.* **28**:287-294 (1991).
26. Z. Moldoveanu, M. Novak, W. Huang, R.M. Gilley, J.K. Staas, D. Schafer, R.W. Compans and J. Mestecky, Oral immunization with influenza virus in biodegradable microspheres, *J. Infect. Dis.* **167**:84-90 (1993).
27. D.T. O'Hagan, J.P. McGee, J. Holmgren, A.MCl Mowat, A.M. Donachie, K.H.G. Mills, W. Gaisford, D. Rahman and S.J. Challacombe, Biodegradable microparticles for oral immunization, *Vaccine* **11**:149-154 (1993).
28. R. Edelman, R.G. Russell, G. Losonsky, B.D. Tall, C.O. Tacker, M.M. Levine and D.H. Lewis, Immunization of rabbits with enterotoxigenic *E. coli* colonization factor antigen (CFA/I) encapsulated in biodegradable microspheres of poly(lactide-co-glycolide), *Vaccine* **11**:155-158 (1993).
29. J.-M. Grzych, D. Grezel, C.B. Xu, J.-L. Neyrinck, M. Capron, J.H. Ouma, A.E. Butterworth and A. Capron, IgA antibodies to aprotective antigen in human schistosomiasis mansoni, *J. Immunol.* **150**:527-535 (1993).

POLYMERIC MATERIALS IN WOUND HEALING

Kezban Ulubayram and Nesrin Hasırcı

Middle East Technical University
Department of Chemistry
06531 Ankara, TURKEY

INTRODUCTION

There is a high need for rapid healing in the treatment of severe burns, trauma, diabetic, decubitus ulcers and other conditions where a great damage of the tissue is exist. In those cases, wound should be covered with a dressing which replaces the functions of the natural skin by protecting the loss of body fluid and proteins, preventing bacterial invasion, improving and stimulating the healing process by providing a support for the proliferating cells.

The principle function of a wound dressing is to provide an optimum medium for the proliferating cells. At different stages of the healing process, major changes occur in the wound environment. These dramatic changes may be represented by the following triptych. In the first act of the repair, the acute inflammatory events limit damage and clear the stage for subsequent repair to take place. Wounds must be protected from further damage, infection must be controlled and debris must be cleared. In the second act, or proliferative phase, formation of fibrovascular granulation tissue and epithelization occur. An optimal wound environment must be provided to allow rapid repair and regeneration. The third act is remodelling and maturation of scar tissue.

Effective wound management requires an understanding of the process of tissue repair and a knowledge of the properties of the dressings available. Only when these two factors are considered the selection of dressing can be done in a logical and informed way.

Wound Dressings

The ability of the human body to heal its wounds involves its more primitive function of preserving life, and there are many factors both favorable and unfavorable that can influence the series of well ordered cellular and biochemical events in the repair process.

Generally, most wounds heal naturally without complex intervention, even in the presence of a limited degree of infection. However, various substances have been used to cover the wounds from the ancient times[1]. The Ebers Papyrus (c.1550 BC) accepted the oldest complete Egyptian medical treatise (castor oil) for application to septic wounds and burns and the use of hartshorn "to drive out painful swellings" [2]. Absorbent cotton wool was used for dressing by Gamgee in 1880 and "tulle gras" by Lumiere in 1903. At that time the great majority of skin wounds, including burns and ulcers, have been dressed with soft paraffin impregnated gauze (tulle gras) and absorbent cotton wool supported by a bandage. Antiseptic and bacteriostatic medications have been topically applied to the wound bed prior to dressing. Later on, the use of synthetic and natural polymers became more popular.

Synthetic polymeric wound dressings are generally thin layer of polymers made of silicone, polyurethane[3,4], polyvinyl chloride or polyethylene. These are flexible and impermeable to bacteria. However, they have a low water-vapor transmission and lead to pooling of exudate under the dressing; for the reason they do not adhere to the wound bed. Most of these films have a synthetic adhesive as a coating on the inner surface that adheres well to dry skin at the wound margins but does not adhere to the wound site. The fact that the film adheres well to the edges of normal skin creates an enclosed environment that protects the wound against bacterial invasion. In addition, antibacterial agents (such as iodine) can be incorporated into the polymeric membrane. Polymeric wound dressings are used only on small shallow wounds that can heal by epithelial cell movement across the wound from the wound edges.

A variety of natural polymers including collagen, fibrin, fibronectin, alginate and hyaluronic acid have been studied as dressings for dermal wounds. Unlike synthetic polymers, natural materials act as inert coverings for the wound.

Most of the wound dressings have been made by combining two or more materials as layers. The outer layer is used as epidermis and designed for durability, elasticity and eventually suturability to wound edges. The inner layer is designed for maximum adherence and elasticity and replaces the function of the dermis. These type of composite materials generally consist of a fabric mesh (cotton gauze, dacron flocking) or biological polymers (collagen etc.) bonded to synthetic polymer films (polyurethane and polysilicones etc.).

Biobrane® is composed of a silicone rubber layer attached to a knitted nylon fabric that also contains porcine collagen-derived peptides. The biological component is believed to increase adhesion to the wound and attract new connective tissue cells. Epigard® consists of an inner layer of reticulum polyurethane foam laminated to an outer sheet of microporous polypropylene film. It adheres firmly to the wound and is stripped from the wound when it is replaced. Some other composite dressings use polytetrafluoroethylene or polyurethane films as the outer layer and pectin or methycellulose as inner adhesive layer. Methylcellulose is more adherent to the wound and impermeable to bacteria and water.

A bilayer composite wound dressing was developed by Yannas and coworkers at Massachusetts Institute of Technology in 1980 and consisted of a silicone membrane attached to an inner layer of collagen sponge[5-10]. The collagen sponge was formed in the presence of chondroitin-6-sulfate and crosslinking was achieved using glutaraldehyde. The silicon layer prevents excessive water-vapor transmission and bacterial invasion and can be sutured onto the wound edges. This composite matrix (collagen / chondroitin-6-sulfate) having a well-described pore structure and crosslinking density that optimize regrowth of the cells while minimizing scar formation[5,8]. Similar bilayer wound dressings (in structure) were developed by Suzuki et al[11] and Matsuda and co-workers modifying Yannas's technique[12,13]. Table 1 classifies some commercially available dressings based on composition, physical properties and usage[14].

Wound Healing and Dressings

The healing process of a wound is a complex process, involving a number of interrelated reactions such as clotting, inflammation, cell migration, tissue replacement and fibrosis. Most dressing research has been directed towards providing the ideal environment for these processes to occur naturally. Parallel to this the research on wound dressing which give increased healing rates may not only give a better quality of wound repair but may also reduce hospitalization times. One adverse effect of commonly used dressings is to cause fresh damage to the wound during removal because of adherence to the wound surface. This adhesion is caused by changes in the properties of the proteinaceous exudate which penetrates among fibres and into any surface irregularities of dressings.

Table 1. Examples of some commercially available dressings

Type	Name / Manufacturer	Description
Adherent, Absorbent, Nonocclusive		Many absorbent woven and nonwoven gauze products
Adherent, Nonabsorbent, Nonocclusive	Adaptic / Johnson&Johnson Inc.	Knitted cellulose acetate impregnated with petrolatum emulsion
	N-Terface / Winfield	Polyethylene-based mesh
Nonadherent, Nonabsorbent, Nonocclusive	Melonin / Smith & Nephew	Nonadherent perforated poly-(ethylene terephtalate) film backed by absorbent cellulosic acrylic layer
	Silicone NA / Johnson&Johnson Inc.	Knitted cellulose acetate fabric with silicone coating
	Sorbsan / Steriseal	Calcium alginate material
	Telfa / Kendal	Cotton sandwich between perforated nonadherent PETF sheets
Nonadherent, Absorbent, Occlusive, Semiocclusive		
Hydrocolloid Dressing	Comfeel Ulcus / Coloplast	Absorbent carboxymethyl cellulose adhesive layer backed by a polyurethane film
	Duoderm / Granuflex in UK	Hydrocolloid layer composed of gelatin, pectin, sodium carboxymethycellulose and polyisobutylene, backed by hydrophobic foam
Composite Dressing	Viasorb / Sherwood	Cotton polyester pad contained within a polyurethane sleeve
Hydrogel Dressing	Geliperm / Gersthtch-Pharma Pougera	Hydrogel of polyacrylamide and agar
	Vigilon / Bard	Crosslinked polyethylene oxide hydrogel (95% water) between two polyethylene films
Semiocclusive, Occlusive, Nonabsorbent	Biocclusive / Johnson & Johnson Medical Inc.	Transparent polyurethane film with acrylic adhesive
	Opsite / Smith & Nephew	Polyurethane film with polyether adhesive
Biological Dressing	Biobrane / Woodreat laboratories	Silicone-nylon collagen bilayer composite
	E Z Derm / Genetic Labs	Porcine xenograft, crosslinked collagen
Medicated Dressing	Bactigras / Smith & Nephew	Chlorhexidine tulle gras
	Tegaderm Plus / 3M	Contains idophor
Hemostasts, Absorbable	Collagen hemostat include Avitene, Hehstat, Oxycell etc.	

As it dries it becomes a powerful glue, binding the dressing into the scab. Removing of adherent dressing causes pain and creates fresh damage to the wound. The practical solution is to prevent dehydration of the exudate at the interface between the dressing and the wound. The traditional and widely used dressing fabrics such as lints and gauzes have high absorbency but also high adhesion to the wound. Therefore fabric dressings absorb readily, even fiercely and the blood and exudate spread far along the fibres. A fabric thoroughly incorporated into scab can not be removed without damage to the healing tissues. Cotton gauze although provides a relatively poor environment for healing, the popularity is partly a matter of tradition, partly a matter of convenience.

In order to provide absorbency, a number of dressings were designed. Non-adherent and absorbent dressings are generally obtained by covering absorbent pad with nonadhering perforated films. Examples of this type of dressings are Melonin®, Telfa®, Perfron® and Lotus®. Melonin® comprises an absorbent cotton and gauze pad faced on one side only with a perforated polyester film. Telfa® is essentially the same, the main differences being that the pad is faced on both sides by the polyester film and the perforations are larger. These dressings have been studied in controlled animal trials using porcine skin and the standard shallow wound models[15] and similar pattern of repair was observed for both, with areas of dehydration and leucocytic concentrations, directly related in size and position to the film perforations, being seen on the wound surface at 3 days. Removel of the Telfa® dressings caused more extensive epidermal loss than removal the Melonin®, although evidence of fresh bleeding was seen with both of dressings. No adverse effects were noted beneath either of these two dressings. Perfron® and Lotus® are similar to each other in composition. Perfron® dressing are absorbent cotton pads in a sleeve of apertured viscose non-woven fabric coated with polypropylene, while in Lotus® the cotton pad is covered by an apertured non-woven fabric comprising 90 % polypropylene and 10 % polyamide. There are similarities in the healing patterns beneath these two dressings when compared with Melonin®, Telfa®. For instance, the wound surface had remained moist, dehydrated leucocytes and dermal collagen can be seen in relation to the sites of apertures. The effects of removing both Perfron® and Lotus® pads from partial thickness wounds were comparable. The wounds were denuded of all moist exudate, but the areas of dehydration remained intact on the wound surface. Little epidermal loss was recorded but dressing removal resulted in the rupture of the dermo-epidermal junction, and once the epidermis has been lifted from the underlying dermis it is unlikely to remain viable.

In the last twenty years, development of alternative dressings have been improved[16]. These new polymer-based dressings include the semi-occlusive films (e.g., OpSite®, Bioclusive® and Steridrape®) and hydrogels (e.g., Vigilon® and Geliperm®) and the occlusive hydrocolloids (e.g., Comfeel Ulcus® and Granuflex®).

Semi-occlusive films were originally designed for use as adhesive surgical drapes and also as dressings. They are composed of thin polyurethane film spread on one surface with an adhesive which gives adhesion to the dry skin and non-adhesion to a wet surface. They are highly comformable due to elastomeric and extensible properties. They are impermeable to water and bacteria but permeable to varying degrees to gaseous transfer. Because these dressings do not incorporate an absorbent component, exudate from the wound collects beneath the dressing to form a bag of fluid. Using OpSite® on the standard partial-thickness pig wound it was found that the volume of retained fluid reach a maximum at 3 days and if dressings were left undisturbed this had reduced at 5 days, by evaporation through the film to leave a thin layer of gelled exudate on the wound surface. The elasticity of the film allowed it to contract back to conform to the contours of the wound. Buchan et al. studies the wound exudate from human graft donor sites collected under OpSite® and compared it to the exudate from pig wounds collected under similar conditions[17]. They demonstrated that the wound exudate, under OpSite® contains large numbers of actively bactericidal neutrophils, high levels of lysozyme and clinically normal levels of plasma proteins[18]. The exudate is therefore actively bactericidal and should be left undisturbed unless clinical signs of infection indicate the need for topical antibacterial agents.

Hydrogels are also semi-occlusive and have the absorbent capacity not provided by the semi-permeable films. They are composed of insoluble hydrophilic polymers arranged in three dimensional network. They are prepared from a variety of materials including gelatin, polysaccharides, polyacrylamide polymers and polymers derived from methacrylic esters. Hydrogels currently available include Geliperm® and Vigilon®. Geliperm® is an elastic gel-film composed of 96 % sterile water bound by a polyacrylamide and agar network[19]. Vigilon® is a colloid in gelatinous form with 96 % water as the dispersion medium and 4 % insoluble crosslinked polethylene oxide. It is centered by a low density polyethylene net to provide strength[20]. Because of high water content of these materials and their high gaseous permeability, they can rapidly dry out if not protected. Secondary dressings of an absorptive pad and / or bandage or periodic re-wetting of dressing may be recommended. Vigilon® dressings are faced with an inert polyethylene film to control water vapor transmission. The histology and standard partial-thickness wounds covered with Vigilon® and Geliperm® show epidermal regeneration occurring within a moist exudate directly over the dermal surface. There has been no dehydration or dermal involvement[21].

Occlusive hydrocolloids are compounds containing hydrogels together with elastomeric and adhesive components. They are completely occlusive, the exclusive of atmospheric oxygen being to encourage the development of a well vascularized wound bed. Hydrocolloid dressings currently available include Granuflex®, Dermiflex®, Comfeel® and Tegasorb®. They are based on sodium carboxymethycellulose, which acts as the primary gelling element and absorbent dressing, to which has been added elastomers and other additives. This inner adhesive layer is covered with an impermeable foam outer layer. Like hydrogels, hydrocolloids will swell in the presence of fluid, but unlike hydrogels this is not a three-dimensional response of the entire gel. Hydrocolloids swell in a linear fashion with a higher moisture retention at the wound / dressing interface. The gel expands proportional to the amount of exudate available and so fills the wound defect. The presence of the outer foam layer means that a pressure is maintained on the floor of the wound by the swollen gel. Hydrogels remain chemically inert in the presence of wound fluids but hydrocolloids interreact with the exudate. In the case of the Granuflex®, which contains a polymer mix of polyisobutylene, gelatin, and pectin, the gels degrades to release the available protein and polysaccharides. The resulting colloidal gel absorbs the soluble components from the exudate and also removes bacteria and cellular debris since the white cells remain viable and capable of phagocytosis. Two strains of bacteria that Granuflex® is claimed to be very effective against are Staph. aureus and Pseudomonas aeruginosa[22]. The mode action of hydrocolloid dressings makes them particularly suitable for the treatment of ulcers.

Biological materials are also used in the production of natural dressings. Some of these materials which have been studied include sphagnum moss[23,24], sugar[25] and calcium alginate extracted from seaweed[26]. Sphagnum has far more medical importance than other mosses mainly because of its great absorbent power and its slight antiseptic properties and it has a long history of use in wound treatment. Sugar and sugar related compounds such as honey have been used in the treatment of wounds for thousands of years. Two sugar paste formulations have been developed and both have excellent in vitro antimicrobial activity and have provided effective treatment for infected and malodorous wounds[27,28]. Alginates, derived from alginic acids of seaweed, are highly absorbent, gel-forming materials with haemostatic properties. Calcium alginate has a well established history of use in wound dressings, principally in the form of woven or knitted fabrics[29]. More recent developments in the medical uses of alginates have resulted in a non-woven calcium alginate for use as a primary dressing. In contact with the body fluids, alginates are known to break down to simple monosacharide type residues and be totally absorbed. The wound exudate converts the calcium to the sodium salt facilitating the removal of the dressing by dissolution. Any residual fibres remaining within the wounds are biodegradable thus eliminating the need for complete removal.

Which Dressings For Which Wound

The process of dressing selection is determined by a number of factors including the nature and location of the wound and the range of materials available. In most situations the cost of treatment is also a major factor. Depending upon their structure and composition, dressings may variously be used to absorb exudate, combat odour and infection, relieve pain, promote autolytic debridement (wound cleansing) or provide and maintain a moist environment at the wound surface to facilitate the production of granulation tissue and the process of epithelialisation. Some dressings simply absorb exudate or wound fluid and may therefore be suitable for application to a variety of different wound types. Others have a very clearly defined specialist function and as such have a more limited range of indications. The purpose or principal aim of the proposed treatment should be defined properly before selecting the dressing material. In most instances this is the healing in the shortest possible time. Occasionally, the removal or containment of both the odour and the exudate is important such as in case of malignant wounds[30].

A simple wound classification system such as that shown below forms a useful starting point in the selection process. Within this classification wounds are divided into four basic types according to their appearance,[31]

1) Necrotic wounds - covered with devitalised epidermis, frequently black in colour.

2) Sloughy wounds - contain a layer of viscous adherent slough, generally yellow in colour.

3) Granulating wounds - contain significant amounts of highly vascularised granulation tissue, generally red or deep pink in colour.

4) Epithelialising wounds - which show evidence of a pink margin to the wound or isolated pink islands on the surface.

It will be recognised that these descriptions relate not only to different types of wounds but also to the various stages through which a single wound may pass as it heals.

For necrotic wounds there are various applications. The first, and probably the least efficient, is by the application of soaks or wet packs consists of gauze pads or nonwoven swabs soaked in water, saline or other solutions such as sodium hypochlorite (Dakin's solutions or Eusol). A more convenient method is the application of a hydrogel dressing such as Intrasite Gel (Smith and Nephew Medical Ltd), Granugel (Convatec Ltd), Sterigel (Seton Healthcare Ltd) Nu-gel (Johnson and Johnson Medical Ltd), etc.. The gel is placed on the wound and covered with an appropriate secondary dressing such as Melolin or Telfa or a vapour permeable film such as Opsite, Tegaderm or Bioclusive. Some products, but not all, are also able to absorb a limited amount of fluid from exuding wounds[32]. An alternative method of rehydrating necrotic tissue depends upon the use of a hydrocolloid dressing. These are available from many different manufacturers but as with the hydrogels, despite superficial similarities in appearance, significant differences exist between the different brands[33].

For sloughy wounds traditionally, agents such as sodium hypochlorite and hydrogen peroxide have been used in the form of soaks but these are of limited efficacy and may also have an adverse effect upon the healing process[34]. Aserbine, a proprietary solution containing malic acid, benzoic acid and salicylic acid is also available but this is not widely used by wound care specialists. Probably the first modern dressing to be marketed specifically for use as a wound cleansing agent was the polysaccharide bead dressing Debrisan (Pharmacia Ltd). When applied to relatively small moist sloughy wounds the beads absorb fluid and progressively move bacteria and cellular debris away from the surface of the wound[35]

Iodosorb (Perstorp) although similar to Debrisan in appearance, also contains elemental iodine which is liberated to exert an antibacterial effect in the wound when the dressing absorbs liquid. Both Debrisan and Iodosorb are also produced in the form of pastes or ointments. Sloughy wounds which also produce a degree of exudate may be dressed with alginate dressings such as Sorbsan (Maersk), Tegagen (3M Health Care Ltd), Kaltostat (Convatec Ltd) or other gel forming polysaccharide dressings such as Aquacel (Convatec Ltd). Polysaccharides such as honey and sucrose have also been used to facilitate wound cleansing. Although ordinary granulated or icing sugar has been used successfully,[36,37] recent interest has been focused on the use of a sugar paste containing polyethylene glycol 400 and hydrogen peroxide[38-40]. Whichever technique is selected, once the slough has been removed, the formation of granulation tissue can take place unhindered.

Granulating wounds vary considerably in size, shape, and the amount of exudate that they produce. As a result, no one dressing will be suitable for use in all situations. Cavity wounds, traditionally packed with gauze soaked in saline, hypochlorite, or proflavine, are now more commonly dressed with alginate fibre in the form of ribbon or rope. In the past, larger cavities were managed very successfully with a silicone foam dressing formed in situ from two liquids carefully measured out in the correct proportions and mixed thoroughly before being introduced into the wound[41]. Cavi-Care can be introduced into large cavities. An alternative dressing for cavity wounds, Allevyn Cavity Wound Dressing (Smith & Nephew Medical), consists of foam chips enclosed in a soft flexible plastic pouch the surface of which contains small perforations to allow the entry of exudate. For more shallow heavily exuding wounds such as leg ulcers, fibrous sheet dressings made from alginate fibre are commonly used. If exudate production is not a problem, the use of a hydrocolloid dressing may be preferred. Other dressings which are used for the treatment of chronic exuding wounds include the highly absorbent hydrophilic foam products Allevyn and Allevyn Adhesive (Smith & Nephew Medical Ltd), an island dressing Tielle, (Johnson and Johnson), and Lyfoam Extra (Seton Healthcare). Recently launched Combiderm (Convatec Ltd), which consists of a self-adhesive absorbent pad containing a superabsorbent in powder form, is also capable of absorbing and retaining large volumes of fluid even under pressure. For more lighly exuding wounds, thin polyurethane foam products are available such as Lyofoam (Seton Healthcare), which has limited absorbency but which is highly permeable to moisture vapour. Lyofoam is also used as low-adherent dressing for minor injuries and other wounds in the final stages of healing. Products containing an antibacterial agent may be used in conjunction with systemic antibiotic therapy, to control the infection which causes the problem, and dressings containing activated charcoal such as Actisorb Plus (Johnson and Johnson Medical Ltd) or Lyofoam C (Seton Healthcare Ltd), can be applied to control the odour.

For epithelialising wounds, fluid production can be a problem in case of burns. Traditionally, these wounds have been dressed with paraffin gauze covered with a layer of Gauze and Cotton Tissue ("Gamgee") but some centres have reported that both alginates[42] and hydrocolloid dressings[43] offer significant advantages in these situations. Other dressings which are used in the final stages of the healing process include the perforated plastic film dressings and the knitted viscose products such as N-A Dressing and Tricotex (Smith & Nephew Medical Ltd). Recently three new low adherent wound contact layers have been introduced. These are NA Ultra (Johnson and Johnson Medical Ltd), which consists of a knitted viscose fabric impregnated with silicone, Mepitel (Molnlycke) which also consists of a knitted fabric impregnated with silicone, and Tegapore (3M Health Care Ltd), a thin sheet of a woven nylon net with very small, well defined holes.

The choice of the dressings material depend on the wound type as well as the properties of the product and situation of the patient. The product related factors may therefore be summarised as follows; conformability, mass or volume (for cavity wounds), fluid handling properties, sensitisation potential, odour absorbing properties, antibacterial activity, haemostatic properties, permeability to tissue fluid and microorganisms, ease of use, pain related factors, fibre-fast, non toxic, cost and availability. The major patient related factors may be summarised as follows; wound aetiology, state of continence, known sensitivity to medicated dressings, fragile or easily damaged skin, the need to bathe or shower frequently and compliance.

At the moment, there is no one dressing satisfying all the desired requirements. However succesful wound management is possible by having a good knowledge of the wound types and a good selection of available dressings.

REFERENCES

1. A. O. Whipple. *The Story of Healing and Wound Repair*, C. Thomas, ed., Springfield, (1963).
2. D. Gutrie, *History of Medicine*, London, (1964).
3. K. Ulubayram and N. Hasırcı, Polyurethanes: effect of chemical composition on mechanical properties and oxygen permeability, *Polymer*, 33(10): 2084-2088 (1992).
4. K. Ulubayram and N. Hasırcı, Properties of plasma modified polyurethane surfaces, *J. Coll and surf B: Biointeractions*, 1:261-269 (1993).
5. I. V. Yannas and J. F. Burke, Design of an artifical skin. I. Basic design principles, *J. Biomed Mater Res* 14:65-81 (1980).
6. I. V. Yannas, J. F. Burke and M. Warpehoski, Prompt, long-term functional replacement of skin, *Trans Am Soc Artif. Intern Organs*, 27:19-23 (1981).
7. J. F. Burke, I. V. Yannas and W. C. Quinby, Successful use of physiologically acceptable artifical skin in the treatment of extensive burn injury, *Ann.Surg*, 194:413-428 (1981).
8. I. V. Yannas, J. F. Burke, P. L. Gordon, C. Huang and R. H. Rubenstein, Design of an artificial skin II. Control of chemical composition, *J Biomed. Mater Res*, 14:107 (1980).
9. I. V. Yannas, J. F. Burke, D. P. Origill and E. M. Skraubut, Wound tissue can utilize a polymeric template to synthesize a functional extension of skin, *Science*, 215:174-176 (1982).
10. I. V. Yannas, What criteria should be used for designning artificial skin replacement and how well do the current grafting materials meet these criteria, *J Trauma*, 24:29-39 (1984).
11. S. Suzuki, K. Matsuda, N. Isshiki, Y. Tamada and Y. Ikada , Experimental study of newly developed bilayer artifical skin, *Biomaterials*, 11:356-360 (1990).
12. K. Matsuda, S. Suzuki, N. Isshiki, K. Yoshioka, R. Wada, S. H. Hyon and Y. Ikada, Evaluation of a bilayer artifical skin capable of sustained release of an antibiotic, *Biomaterials,* 13:2 (1992).
13. K. Matsuda, S. Suzuki, N. Isshiki, K. Yoshioka, R. Wada, T. Okada and Y. Ikada, Influence of glycosaminoglycans on the collagen sponges component of a bilayer artificial skin, *Biomaterials*, 11:351-355 (1990).
14. D. M. Wiseman, D.t. Rovee and O. M. Alvarez, Wound dressings: design and use, in Wound Healing: Biochemical and Clinical Aspects, J. Micheil, ed., W. B. Saunders Comp., Philadelphia, (1992).
15. S. J. Varley, S. E. Barnet, A study of wound dressing adhesion, *Clinical Materials*, 1:37-57 (1986).
16. T. D. Turner, Semiocclusive and occlusive dressings, in: Enviroment for Healing:The Role of Occlusion, International Congress and Symposium Series. 88, Royal Society of Medicine, London (1985).
17. I. A. Buchan, J. K. Andrews, S. M. Lang, J. G. Boorman, J. V. K. Harvey and B. G. H. Lamberty, Clinical and laboratory investigation of the composition and properties of human skin wound exudate under semi-permeable dressings, *Burns*, 7:326-334 (1980).
18. A. Buchan, J. K. Andrews, S. M. Lang, Laboratory investigation of the composition and properties of pig skin wound exudate under Op-Site, *Burns*, 8:39-46 (1981).
19. Trade Literature, Geistlich-Pharma, Switzerland.
20. Trade Literature, Bard Ltd., Sunderlan, England.
21. S. E. Barnet and S. J. Irving, Studies of wound healing and the effects of dressings, in: High Performance Biomaterials, M. Szycher, ed., Technomic Publishing Comp., Lancester, USA (1991).
22. Trade Literature, Squibb Surgicare Ltd., Hounslow, England.
23. J. Varley, S. E. Barnett, Sphagnum moss and wound healing I, *Clinical Rehabilitation*, 1:147-152 (1987).
24. J. Varley, S. E. Barnett, Sphagnum moss and wound healing II, *Clinical Rehabilitation*, 1:153-160 (1987).
25. H. G. Archer, S. E. Barnett, S. J. Irving, K. R. Midleton and D. V. Seal, A controlled model of moist

wound healing: comparison between semi-permeable film, antiseptic and sugar paste, *J Exp. Path* 83:17,(1989).

26. S. E. Barnett and S. J. Varley, The effects of calcium alginate on wound healing, *Ann Royal, Coll, Surg, Eng*, 69:153-155 (1987).

27. K. Middleton and D. V. Seal, Sugar as an aid to wound healing, *Pharm J*, 235:757-758 (1985).

28. K. Middleton and D. V. Seal, Development of a semi-synthetic sugar paste for promoting of infected wounds, in: Pathogenesis of Wound and Biomaterial-Associated Infections, I. Eliasson,ed., London, England, Springer-Verlag (1989).

29. G. Blaine, Experimental observation on absorbable alginate products in surgery, *Ann Surg*, 125:102-107 (1947).

30. S. Thomas, Treating malodorous wounds, *Community Outlook*, Oct, 27-29 (1989).

31. S. Thomas, http://www.smtl.co.uk./Wourd-Wide-Wounds/1997/july/Thomas-Guide/ Dress -Select. Html.

32. S. Thomas and N. P.Hay, In vitro investigations of a new hydrogel dressing, *Journal of Wound Care*, 5(3):130-131 (1996).

33. S. Thomas and P. Loveless, An examination of the properties of 12 hydrocolloid dressings, World wide wounds, http://www.smtl.co.uk/World-Wide-Wounds/1997/july/ Thomas-Hydronet /hydronet. Html.

34. D. Moore, Hypochlorites: A review of the evidence, *Journal of Wound Care*, 1(4):44-53 (1992).

35. S. Jacobsson , A new principle for the cleansing of infected wounds, *Scand J Plast Reconstr Surg* 10:65-72 (1976).

36. Sugar sweetens the lot of patients with bedsores, *JAMA*, 223, 122 (1973).

37. R. A. Knutson, Use of sugar and povidone iodine to enhance wound healing: five years experience, Sth Med J, 74:1329-1335 (1981).

38. H. Gordon, Sugar and wound healing, *Lancet*, 2:663-664 (1985).

39. K. R. Middleton and D. Seal, Sugar as an aid to wound healing, *Pharm J*, 235:757-758 (1985).

40. H. G. Archer, A controlled model of moist wound healing: comparison between semipermeable film antiseptics and sugar paste, *J exp Path*, 75:155-170 (1990).

41. R. A. B. Wood and L.E. Hughes, Silicone foam sponge for pilonidal sinus: a new technique for dressing open granulating wounds, *Br Med J*, 3:131-133 (1975).

42. A. R. Groves and J.C. Lawrence, Alginate dressings as a donor site haemostat, *Ann R Coll Surg*, 68:27-28 (1986).

43. Doherty, Granuflex hydrocolloid as a donor site dressing, Care of the Critically Ill, 2:193-194 (1986).

EPIDERMAL GROWTH FACTOR (EGF) WOUND HEALING IN FLUOROCARBON AND CHITOSAN GELS IN A RABBIT MODEL

Erem Memişoğlu[1], Filiz Öner[2], H.Süheyla Kaş[1], Leila Zarif[3], Ayşe Ayhan[4], İhsan Başaran[5], and A.Atilla Hıncal[1]

[1]Hacettepe University
Faculty of Pharmacy
Department of Pharmaceutical Technology
06100 Ankara, TURKEY
[2]Hacettepe University
Faculty of Pharmacy
Department of Pharmaceutical Biotechnology
06100 Ankara, TURKEY
[3]BioDelivery Sciences Inc.
Ave du Groupe Morgan
06700 St. Laurent du Var, FRANCE
[4]Hacettepe University
Faculty of Medicine
Department of Pathology
06100 Ankara, TURKEY
[5]Hacettepe University
Faculty of Medicine
Department of Plastic and Reconstructive Surgery
06100 Ankara, TURKEY

INTRODUCTION

Wound repair follows a general scheme, a sequence of processes taking place in an orderly way: inflammation, repair and closure, remodelling and final healing. Growth factors are produced by the cells aiding the process and are effective during replacement and reconstitution [1]. A wound is defined as an interruption of tissue to a greater or lesser extent, which may affect skin, mucosa or organs. The specific sequence of different processes following wounding has one common aim: repair. In every wound type, the healing process runs through three stages, which partly overlap. The first one, the exsudative or inflammatory phase, is followed by the proliferative phase and finally the regenerative phase.

Biomedical Science and Technology
Edited by Hıncal and Kas, Plenum Press, New York, 1998

Characteristic for the inflammatory phase, lasting approximately 72 h, is the activation of blood coagulation system and the release of various mediators from platelets. This is followed by the coagulation of blood, within 2-4 h inflammatory cell immigration starts and after 32 h fibroblasts are present in the wound site. The second phase of wound repair is characterized by proliferation and lasts from day 1 to a maximum of 14 days. Highly vascularized granulation tissue is formed and angiogenesis and neovascularization in the wound site starts. During the last phase of wound healing the production of new connective tissue is of main importance. If all epidermal layers are effected, re-epithelialization proceeds through the following three stages: migration of basal lamina cells, mitosis of cells migrating across the wound surface and maturation of newly generated cells. The final step in epidermal wound healing is characterized by cell maturation, leading to the regeneration of a defined epidermal layer. Keratinization starts and finally desmosomes promote attachment of cells to one another. The wound is closed and covered by mature epidermis [2].

Epidermal Growth Factor (EGF) is a 53 amino acid, single-chain polypeptide produced in the salivary glands and is known to stimulate epidermal cell proliferation. EGF was reported to accelerate closure of full-thickness skin injuries to rabbit ears. It was examined in a polyvinyl sponge model and a hollow sponge model and found to accelerate the rate of cell and collagen accumulation both in vitro and in vivo. Recently, it has been shown to increase the rate of epithelialization in human skin graft donor sites [3].

The ordered deposition of collagen and the histoarchitecturally valid reconstruction of the cutaneous wound tissues can be modulated by N-acetyl glucosamine polymers supplied to the wound[4]. N-carboxybutyl was reported to favour neoangiogenesis and prevented the regression that the wound usually undergoes from capillary-rich to avascular scar composed of dense collagen bundles [5].

Fluorocarbons are in clinical trials as oxygen carriers. These compounds have been known for years for their high capacity to dissolve gases, particularly oxygen and carbon dioxide. When emulsified, fluorocarbons can be used as blood substitutes. The fluorocarbon chain can constitute the hidrophobic tail of amphiphiles. A large number of well-defined neutral, anionic, zwitterionic fluorocarbon amphiphiles were synthetized to be used as components for colloidal systems. The presence of fluorine atoms was found to allow the formation of stable gels. When the fluorine atoms are present in sufficient amount, biocompatible, highly surface active and non hemolytic amphiphiles are obtained. Stable, heat-sterilizable fluorocarbon gels can be obtained with fluorinated telomers [6]

The objective of this study was to evaluate the enhancing effect of chitosan and fluorocarbon gels as topical vehicles for the wound healing effectiveness of active ingredient EGF. The treatment groups in this study were EGF in chitosan gel, EGF in fluorocarbon gel, EGF in aqueous solution, empty chitosan gel, empty fluorocarbon gel and the control group was kept untreated.

MATERIAL AND METHODS

Male rabbits used in this study ranged in weight from 8 to 12 kg and were housed for two weeks prior to wounding in proper conditions and diet. EGF was a kind gift from Dr. Ali Kemaloğlu (Alke Cosmetics- Turkey), Chitosan-H (Lot 337) was produced by Dainishiseiken Colour and Chemicals MFG Co. Ltd., Japan. Fluorocarbon was supplied by Air Products, France.

1% chitosan gel was prepared by dispersing chitosan-H in 1% (w/v) acetic acid and stirring for 24 h with a magnetic stirrer. EGF was incorporated into the gel formulations as 10 µg/mL with a Hamilton microsyringe.

Wounding and wound care method was modified for the rabbit model [7]. Animals were fasted overnight before wounding and hair on the back of the rabbits were clipped and rinsed with 70% ethanol. Ketamin hidrochloride (Ketalar®, ParkeDavis) and xylazine hidrochloride (Rompun®, Bayer) were injected intramuscularly for anesthesia. Then on each animal, full thickness wounds of 1 cm^2 were formed by excision with a scalpel. These wounds were separately treated with chitosan gel, fluorocarbon gel, EGF+chitosan gel, EGF+fluorocarbon gel, EGF in aqueous solution form and the control group was not treated. Wounds were covered with Op-site semipermeable adhesive dressing after treatment. Wound perimeters were traced onto tracing papers and wound areas were determined on each measurement day. Wounds were considered healed when the moist granulation tissue was no longer visible and wound site was covered by a continuous layer of epithelium.

On the 18th day of the experiment, biopsies covering both the wound site and the healthy tissue surrounding it were taken and fixed in 10% formalin solution for a maximum of 24 h, embedded in paraffin blocks and 6 μm sections were stained with hematoxylin-eosin and evaluated according to the presence of crust, epidermal regeneration, presence of acute inflammatory elements and collagenization of the granulation tissue. Biopsies were also stained with avidin-biotin-peroxidase complex for immunohistochemical evaluation. Antifactor VIII was used to mark vascular epithelium and anticollagen IV was a marker for type IV collagen which represents the formation of organized fibrous tissue in the wound site.

RESULTS AND DISCUSSION

Control group displays an impaired wound healing profile while the medicated groups like EGF+chitosan gel and EGF+fluorocarbon gel display descending wound areas unlike the other groups which display sudden increase and decreases in their profile. This data was analyzed statistically by two-way ANOVA for repeated measures test and the probability values are seen in Table 1.

Table 1. Daily probability values of mean wound areas of treatment groups and control

Groups	day 2	day 4	day 7	day 9	day 11	day 14	day 16	day 18
EGF+chitosan gel vs. EGF solution	p=0.559	p=1.000	p=0.473	p=0.324	p=0.509	p=0.207	p=0.029	p=0.029
EGF+FC gel vs. EGF solution	p=0.457	p=0.146	p=0.368	p=0.408	p=0.396	p=0.210	p=0.129	p=0.135
EGF+FC gel vs. EGF+chitosan gel	p=0.845	p=0.146	p=0.153	p=0.140	p=0.253	p=1.000	p=0.511	p=0.347
FC gel vs. Chitosan gel	p=0.516	p=0.662	p=0.224	p=0.184	p=1.000	p=0.946	p=0.511	p=0.771
EGF+FC gel vs. FC gel	p=0.020	p=0.054	p=0.246	p=0.293	p=0.565	p=0.000	p=0.000	p=0.002
EGF+chitosan gel vs. chitosan gel	p=0.027	p=0.001	p=0.000	p=0.000	p=0.000	p=0.000	p=0.000	p=0.001
FC gel vs. control	p=0.738	p=0.004	p=0.744	p=0.062	p=0.062	p=0.000	p=0.000	p=0.005
Chitosan gel vs. control	p=0.294	p=0.023	p=0.005	p=0.098	p=0.098	p=0.001	p=0.000	p=0.006

Mean wound areas that were measured in certain time intervals are shown in Figure 1.

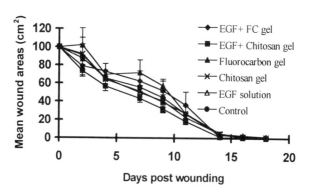

Figure 1. Mean wound areas of all treatment groups vs. control.

A significant difference was observed between EGF+Fluorocarbon gel vs. empty fluorocarbon gel and EGF+chitosan gel vs. empty chitosan gel. These empty gel formulations were also found to be significantly different when compared to the untreated control group. Measurement of wound areas and their statistical evaluation were not sufficient for drawing valid conclusions because the measurement depended highly on hand manipulation. Histological biopsy gradings in Table 2 were analyzed statistically after grading according to four criteria;

1. *presence of crust:* 0: crust present, 1: crust not present
2. *epidermal regeneration:* 0: epidermis non-regenerated, 1: epidermis fully regenerated
3. *presence of acute inflammatory elements (AIE):* 0: AIE present, 1: AIE not present
4. *collagenization of granulation tissue:* 0: Granulation tissue non-collagenized, 1: Granulation tissue collagenized

These gradings were statistically compared by Crosstab/Chi-square test and the probability values in Table 3 revealed that all treatment groups displayed significant difference compared to the untreated control group. This expected effect was seen clearly on the presence of acute inflammatory elements and collagenization of granulation tissue except EGF applied in aqueous solution form which was indifferent than the control group in all criteria. Empty chitosan gel and fluorocarbon gel were also statistically different than the control and the chitosan gel displayed significant difference than the EGF solution probably due to its occlusive effect on the wound. Empty fluorocarbon gel was not as effective since chitosan gel was found to display better healing profile than the fluorocarbon gel in the presence of acute inflammatory elements.

Table 2. Biopsy gradings of all treatment groups and control

Treatment group	Presence of crust	Epidermal regeneration	Presence of AIE	Collagenization of granulation tissue
Control	0,1,1,1,1	0,1,1,1,1	0,0,0,0,0	0,0,0,0,0
Chitosan gel	1,1,1,1,1	1,1,1,1,1	0,0,1,1,1	0,0,1,1,1
EGF+Chitosan gel	1,1,1,1,1	1,1,1,1,1	1,1,1,1,1	1,1,1,1,1
EGF aqueous solution	1,1,1,1,1	0,1,1,1,1	0,0,0,0,0	0,0,0,0,0
Fluorocarbon gel	0,0,1,1,1	0,0,1,1,1	0,0,0,0,0	0,0,0,1,1
EGF+Fluorocarbon gel	1,1,1,1,1	1,1,1,1,1	1,1,1,1,1	1,1,1,1,1

Table 3. Probability values of Crosstab/Chi-square tests performed on the biopsy gradings

Treatment group	Presence of crust	Epidermal regeneration	Presence of AIE	Collagenization of granulation tissue
Control vs. EGF+FC gel	p=0.2918	p=0.2918	p=0.00156	p=0.00156
Control vs. FC gel	p=0.4902	p=0.4902	p=0.2918	p=0.1138
Control vs. Chitosan gel	p=0.1138	p=0.1138	p=0.0384	p=0.0384
Control vs. EGF+Chitosan gel	p=0.2918	p=0.1138	p=0.00156	p=0.00156
Control vs. EGF solution	p=0.2918	p=1.000	p=1.000	p=1.000
EGF+FC gel vs. EGF+Chitosan gel	p=1.000	p=1.000	p=1.000	p=1.000
FC gel vs. Chitosan gel	p=0.1138	p=0.1138	p=0.0384	p=0.1138
EGF+Chitosan gel vs. Chitosan gel	p=1.000	p=1.000	p=0.1138	p=0.1138
EGF+FC gel vs. FC gel	p=0.1138	p=0.1138	p=0.00156	p=0.00156
EGF solution vs. EGF+Chitosan gel	p=0.2918	p=0.2918	p=0.00156	p=0.00156
EGF solution vs. EGF+FC gel	p=0.2918	p=0.2918	p=0.00156	p=0.00156
EGF solution vs. FC gel	p=0.4902	p=0.4902	p=0.2918	p=0.1138
EGF solution vs. Chitosan gel	p=1.000	p=0.2918	p=0.0384	p=0.0384

On the other hand there was no difference between the healing profiles of EGF+fluorocarbon gel and EGF+chitosan gel groups. As our primary objective was the assessment of topical vehicles as enhancers for the wound healing properties of EGF, EGF in aqueous solution form was compared to the fluorocarbon and chitosan gel groups containing EGF. Both formulations displayed optimum healing and were statistically different than both the control and the EGF in solution form.

Histological photomicrographs revealed that EGF+chitosan gel and EGF+fluorocarbon gel groups revealed optimum healing with no crust or acute inflammatory elements, mature epidermis totally regenerated and covering the wound site, granulation tissue was collagenized. As seen in Figure 2, biopsy of EGF+chitosan gel treated group reveals optimum healing as well as biopsy taken from EGF+Fluorocarbon gel treated group which is seen in Figure 3.

On the contrary, control group seen in Figure 4 and EGF in aqueous solution shown in Figure 5 displayed impaired wound healing with excessive acute inflammatory elements, epidermis not regenerated with signs of ulceration and a very vascularized and non-collagenized granulation tissue with necrotized tissue known as crust present on all the wounds of these groups.

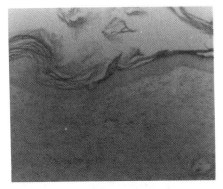

Figure 2. Photomicrograph of EGF+chitosan gel treated group biopsy stained with heamtoxylin-eosin.

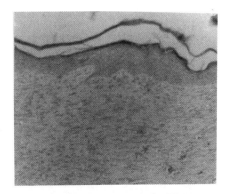

Figure 3. Photomicrograph of EGF+ fluorocarbon gel treated group stained with hematoxylin-eosin.

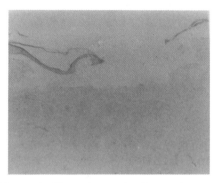

Figure 4. Photomicrograph of control group biopsy stained with hematoxylin-eosin.

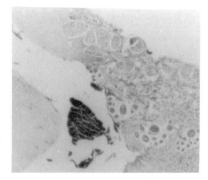

Figure 5. Photomicrograph of EGF in aqueous solution treated group biopsy stained with heamtoxylin-cosin.

Biopsies were also evaluated immunohistochemically by staining with an avidin-biotin-peroxidase complex and to mark vascular endothelium antifactor VIII was used as a marker seen in Figures 6 and 7. Figure 6 shows excessive number of newly branched blood vessels in the control group while Figure 7 displays a little number of neovascularization in an optimally healed wound of EGF+chitosan gel group.

Another immunohistochemical evaluation was performed by marking the type IV collagen in the granulation tissue with anticollagen IV. Figure 8 reveals the very rare collagen bundles with light color in the control group while Figure 9 reveals the dense collagen fibers in the EGF+fluorocarbon gel treated group seen in dark color.

CONCLUSION

In the light of these data, it can be concluded that fluorocarbon gel and chitosan gel are both suitable vehicles for the wound healing agent EGF since these groups have displayed optimum healing while EGF applied in solution form resulted with impaired healing. Due their occlusive effect, these formulations may also function as wound healing agent without any active ingredient but further information and studies on the mechanisn of healing for these formulations will be helpful to draw a valid conclusion from the data obtained in this study.

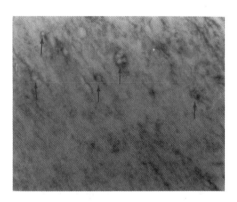

Figure 6. Control biopsy marked with antifactor VIII (➔vascular endothelium).

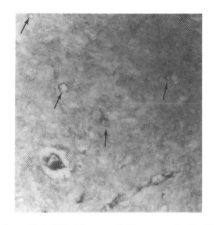

Figure 7. EGF+chitosan gel biopsy marked with antifactor VIII (➔vascular endothelium).

 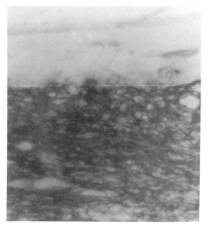

Figure 8. Control group biopsy marked with anticollagen IV (collagen bundles are seen in dark color).

Figure 9. EGF+fluorocarbon gel group biopsy marked with anticollagen IV (collagen bundles are seen in dark color).

Acknowledgement

Authors wish to thank Dr. Ali Kemaloğlu from Alke Cosmetics, Turkey for kindly providing the EGF used in this study.

REFERENCES

1. H. Hammar, Wound healing, *Int. J. Dermatol.*, 32(1); 6 (1993)
2. I.K. Cohen, R.F. Diegelmann, W.J. Lindblad, *Wound Healing: Biochemical and Clinical Aspects*, W.B. Saunders Company, Philadelphia (1992)
3. L.B. Nanney, Epidermal and dermal effects of Epidermal Growth Factor during wound repair, *J. Invest. Dermatol.*, 94; 624 (1990)
4. R. Muzzarelli, V. Baldassarre, F. Conchi, P. Ferrraro, G. Biagini, G. Gazzanelli, V. Vasi, Biological activity of chitosan: an ultrastructural study, *Biomaterials*, 9;247 (1988)
5. G. Biagini, A. Pugnaloni, G. Frongia, G. Gazzanelli, C. Lough, R. Muzzarelli, N-carboxymethyl chitosan induces neovascularization in: *Chitin and Chitosan* G. Skjak-Braek, T. Anthonsen, P. Sandford, ed., Elsevier, London (1988)
6. L. Zarif, J. Riess, A. Pavia, Drug delivery systems based on telomeric fluorocarbon materials, *Proceedings of the 8th International Symposium on Pharmaceutical Technology*, 9-11 September, 1996, Ankara
7. E. Memişoğlu, F.Öner, A. Ayhan, İ. Başaran, A.A. Hıncal, In vivo evaluation for rhGM-CSF wound healing efficacy in topical vehicles, *Pharm. Dev. Tech.*, 2(2); 171 (1997)

USE OF POLYMERS IN DENTISTRY

Saime Şahin

Hacettepe University
Faculty of Dentistry
Department of Prosthodontics
06100 Ankara Turkey

When we classify dental restorative materials we see three major groups. These are metals, ceramics and polymers. Polymers have a major role in most areas of restorative dentistry. The most widely used impression materials are elastomeric polymers. Another major type of polymeric dental material is the composite filling material for anterior teeth. Removable dentures are made from acrylic resin and other polymers. Additional applications as polymers include soft liners, cements, pit and fissures sealant.[1]

IMPRESSION MATERIALS

Impression materials are used to make replicas of oral structures.[2] The function of an impression material is to accurately record the dimensions of oral tissues and their spatial relationships. In making an impression, a material in the plastic state is placed against the oral tissues to set. After setting, the impression is removed from the mouth and is used to make a replica of the oral tissues. The impression gives a negative reproduction of these tissues. A positive reproduction is obtained by pouring dental stone or other suitable material into the impression and allowing it to harden.[3] All impression materials must be in a plastic or fluid state while the replica is being made. After setting, two general classes of impression materials may be distinguished: elastic and nonelastic. Classification of dental impression materials is shown in Table 1. Several polymers (alginates, polysulfides, and silicones) are used as an impression materials.[2]

Alginates

Alginates are the most widely used impression materials in dentistry. They are used for making impressions for partial dentures, complete dentures, orthodontic and study models.[2]

Alginate impression materials contain sodium alginate polymers with fillers (wood flour, etc.) and calcium sulfate.[1] The powder is mixed with water to obtain a paste. Two main reactions occur when the powder reacts with water during setting. First, the sodium phosphate reacts with the calcium sulfate to provide adequate working time. Second, after the sodium phosphate has reacted, the remaining calcium sulfate reacts with the sodium alginate to form an insoluble calcium alginate, which forms a gel with water.[2]

Table 1. Classification of dental impression materials

Elastic materials	Nonelastic materials
Dental compound	Agar hidrocolloid
Impression plaster	Alginate hidrocolloid
Zinc oxide-eugenol	Polysulfide
	Silicone
	Polyether

Polysulfides

Polysulfide rubbers are widely used for crown and bridge application, due to their high accuracy and relatively low cost.[2,3] These materials are useful for multiple impressions when extra time is needed.[2]

The polysulfides are supplied in tubes of base paste and catalyst paste that are mixed together. The base material consists of about 80% low-molecular weight organic polymer, containing reactive mercaptan groups and 20% reinforcing agents, such as titanium dioxide, zinc sulfate, copper carbonate or silica. The accelerator or catalyst paste contains lead dioxide, hydrated copper oxide or cumene hydroperoxide as a catalyst; sulphur as a promoter and dibutylphthalate or other non reactive oils to form a paste.[2,3]

These materials are mixed on a mixing pad with a spatula. Equal lengths of base and catalyst are extruded on a disposable mixing pad. The components are mixed thoroughly with a stiff tapered spatula. Adequate mixing times are 45 to 60 seconds.

Disadvantages include the need to use custom-made rather than stock trays due to greater chance of distortion, a bad odor, a tendency to run down the patient's throat due to lower viscosity, and the lead dioxide materials that stain clothing.[2]

Silicones

Two types of silicones are used as impression materials (1) condensation and (2) addition(vinyl silicone) types. The name identify the type of polimerization reaction that results in the setting of the rubber. Silicone pastes are supplied in light, regular, and heavy body consistencies, as well as a very heavy consistency called a putty. The consistency is controlled by selection of the molecular weight of the dimethylsiloxane and the concentration of reinforcing agent. Higher molecular weights are used with the heavier-bodied materials. The concentration of the reinforcing agent increases from 35% for light-bodied consistency to 75% for the putty consistency.[3]

Condensation Silicone. Condensation silicone impression materials are used mainly for crown and bridge impressions.[2] These materials are supplied either as two paste or paste-liquid catalyst systems. The base paste usually contains a moderately high molecular weight poly(dimethylsiloxane) with terminal hydroxyl groups, orthoalkysilicate for cross-linking and inorganic filler. A paste will contain 30% to 40% filler, where as putty will contain as much as 75%. The catalyst paste or liquid usually contains a metal organic ester, such as tin octoate or dibutyl tin dilaurate and an oily diluent.[2,3]

The manipulation is the same as for polysulfides, except that the silicone material may be supplied as a base paste plus a liquid catalyst. When it is supplied in this form, one drop per inch extruded base paste usually recommended. The setting time (six to eight minutes) is less than that of the polysulfides, which offers some advantage in saving chair time.

The main disadvantages are inaccuracy due to shrinkage on standing and the need for a very dry field, since condensation silicones are hydrophobic.[2]

Addition Silicone (vinyl silicone). Additional silicones represent an advance in accuracy over the condensation silicones. This has been achieved by change in polymerization reactions to an addition type and elimination of an alcohol by product that evaporates, causing shrinkage. Due to their high accuracy, these materials are suitable for crown bridge and partial denture impressions. They are expensive and rigid after setting.[2]

These materials are based on silicone prepolymers with vinyl and hydrogen side groups, which can polymerize by additional polymerization. The setting reaction is produced by mixing one paste containing the vinyl-poly(dimethyksiloxane) prepolymer with a second paste that contains a siloxane prepolymer with hydrogen side groups. A platinum catalyst, which is chloroplatinic acid, is present in one of the pastes and it starts the additional polymerization reaction.[2]

The disadvantages are that it is expensive twice the cost of polysulfides; is more rigid than condensation silicones and is difficult to remove around undercuts: has o low strength, making removal from gingival retraction areas more risky: and may release hydrogen gas on setting and produce bubbles on die surfaces if an absorber is not in the product.[2]

DIRECT ESTHETIC RESTORATIVE MATERIALS

The patient desires esthetic restorations, particularly in the anterior portion of the mouth, and a direct filling material is advantageous in terms of the time required and the cost of the restoration. Four types of materials, silicates (1800), acrylic polymers, dimethacrylate polymers (Composite Resins) (1960), ionomer restoratives (1972) have been developed for use as direct esthetic restorations especially for anterior tooth.[3] Dimethacrylate Polymers (Composite resins) are presently the most popular tooth colored materials having largely replaced silicates and acrylic resin.[3]

Composite Resins

The composite resin was developed by Dr. Raphael Bowen at the National Bureau of Standards / American Dental Association. Compared to acrylic resin, this new polymer was considerably more resistant to wear.[4] Composite resins are commonly used for the restoration of Class 3, 4 and 5 cavity preparations for anterior teeth, veneering of facial or labial surfaces of natural teeth, and more recently for limited restoration of occlusal surfaces. In conjunction with the acid-etch technique, composite resins can be bonded directly to enamel surfaces.[4]

Composite restorations consist mainly of two phases, one a polymer(organic) matrix and the other dispersed filler particles. The polymer consists of an aromatic dimethacrylate monomer (BIS-GMA), which is a reaction product of bisphenol A and glycidyl methacrylate developed by Bowen.[5] The inorganic or filler particles constitute approximately one-half the total volume or 75 to 85 % by weight. Although the composition of this phase varies from product to product, most proprietary materials contain quartz, lithium aluminum silicate, barium glasses, strontium glass, silica or borosilicates. The surface of the particles is coated with an appropriate silane coupling agent to provide adhesion between the filler and the resin matrix. Benzoyl peroxide and a tertiary amine are used as initiators and activators respectively.[3,4,5]

Two principal systems used to achieve polymerization are the chemically activated system and light-activated system. Chemically activated systems usually are supplied as two pastes in two jars, with the initiator in one and the accelerator in the other. The materials most commonly are pigmented to a universal shade that matches the majority of teeth. Additional shades are available, as well as tints that can be mixed with the universal shade, and these provide the dentist with composites to match teeth of other shades. In the light-activated system, the composite is exposed to an intense blue light. The light is absorbed by a diketone which, in the presence of an organic amine, starts the polymerization reaction. Exposure times of 20 to 60 seconds are needed for polymerization. Thus the material is supplied as a single paste in a syringe.[3,6]

Ionomer Restoratives

Clinical studies have shown that the retention of ionomers in areas of cervical erosion are considerably better than for composites.[3] Ionomers are supplied as powders of various shades and a liquid. The powder is an aluminosilicate glass, flour and the liquid is a water solution of polymers and copolymers of acrylic acid. The powder and liquid are dispersed in proper amounts on the paper pad, and half the powder is incorporated to produce a homogeneous milky consistency. The remainder of the powder is added, and total mixing time of 30 to 40 seconds is used with a typical setting time of 4 minutes. After placing the restorative and carving the correct contour, the surface should be protected form saliva by an application of varnish. Trimming and finishing is done after 24 hours. The material sets as a result of the metallic salt bridges between the Al^{+3} and Ca^{+2} ions and the acid groups on the polymers.[3]

DENTURE BASE POLYMERS

Wood, bone, ivory, ceramics, metals, metal alloys and numerous polymers have been used in the denture base applications.[7] Vulcanized rubber or vulcanite was introduced as a denture base material in 1855. This marked the introduction of polymers in complete denture prosthetics.[7] In 1935 Imperial Chemical Industries introduced the first acrylic resin (Kallodent) to the dental profession. Vulcanized rubber eventually was displaced by another polymer, poly(methyl methacrylate) (PMMA) in 1937. During subsequent years, epoxy resin, polystyrene, polyvinyl acrylic, polyamides and polycarbonate have been introduced as denture base materials, but these have only had limited acceptance.[7,8]

A light-activated urethane dimethacrylate also was introduced for denture base applications. None of these materials provided the unique combination of physical and esthetic properties exhibited by PMMA. As a result PMMA has dominated the denture base arena for more than 50 years. Currently, most PMMA denture base materials are supplied as two-component systems.[7,8]

One component is a liquid, whereas the other is a fine, pink powder. The powder contains prepolymerized metil methacrylate PMMA beads, colorants and benzoyl peroxide. PMMA beads serve as important structural components in the polymerized resin. Benzoyl peroxide serves as the initiator for polimerizator. Colorants are provided to simulate the pigmentation of oral soft tissues. The liquid generally contains methyl methacrylate (monomer), glycol dimethacrylate and hidroquinone. Methyl methacrylate provides the building blocks for polymerization, whereas glycol dimethacrylate serves as a crooslinking agent. Hidroquinone is added as an inhibitor. The chemical agent prevent adverting polymerization of the liquid.[3,7,8,9]

When these two components are mixed the mass passes through a series of stages : sandy, stringy, doughy and rubbery. The function of the monomer in the polymer is to produce a plastic mass that can be packed in to the mold. Such a plasticization is accomplished by a partial solution of the polymer in the monomer. At least four stages can be identified during the physical interaction of the powder and liquid:

Stage 1. The powder gradually settles into the monomer to form a some what fluid, incoherent mass.

Stage 2. The monomer attacks the polymer. This is accomplished by the penetration of the monomer into the polymer. This stage is characterised by a stringiness and adhesiveness if the mixture is touched or pulled apart.

Stage 3. As the monomer diffuses into the polymer, and the mass becomes more saturated with polymer solution, it becomes smooth and doughlike. This stage is often called the dough or gel stage. While the mixture is in this stage, it is packed into the mold.

Stage 4. The monomer seemingly disappears, by evaporation and further penetration into polymer. The mass becomes more cohesive and rubberlike.[10]

Several methods are available for processing resin denture bases, such as compression molding (press/pack), injection molding and pouring fluid resins into a vented mold.[10]. Molding of fabricating dentures using PMMA is mostly carried out by the dough molding press/pack technique. In this technique a polymer/monomer mix is forced by the application of pressure to flow into, fill and conform to the shape of the mold.[3]

The denture mold is constructed by using a clinically acceptable waxed-up on a stone cast. The cast, with the record base and positioned teeth, is seated in freshly mixed dental stone or plaster in the appropriate denture flask. The type of flask used is determined by the technique employed in fabricating the denture. When the flask section are separated, the wax and record base are removed. The mold space that remains is filled with the resin denture base material. After it has been formed and polimerized, the denture is removed from the flask and finished, resulting in a final prosthesis.[10]

The polymerization can be effected either by external heating the polymer-monomer mixture, usually in a water bath, by chemical activation at room temperature or by activating the reaction using microwave energy or visible light. The applications of external heating is the most popular system for the denture base polymers, because it offers a variety of applications and because when the process is completed, after the water is allowed to boil, it reduced the residual monomer in the dentures.[11]

SOFT LINERS

Soft lining materials are resilient polymers used to replace the fitting surface of a hard plastic denture, either because the patient can not tolerate a hard fitting surface, or to improve retention of the denture.[9] Soft liners can be classified as long term (Permanent) or tissue conditioner (Temporary) materials. Tissue conditioners or temporary soft liners are materials whose useful function is very short, generally a matter of a few days. Relining the ill-fitting denture with tissue conditioner allows the tissue to return to normal, at which time a new denture can be made.[3,10]

Acrylics and silicones are two main families of polymers used commercially as soft liners.[9] The most common is plasticized acrylic resin, either self curing or heat curing. They are supplied as a powder and liquid. The composition of powder is generally a poly(ethyl methacrylate or one of its copolymers, whereas the liquid is an aromatic ester (butyl phthalate butyl glycolate) in ethanol or an alcohol of high molecular weight.[10]

The silicones used liners can be divided into two types; room temperature vulcanizing (RTV) and heat curing. The heat-cured silicones have in their formulation a siloxane methacrylate that can polymerize into the curing denture base.[9]

CEMENTS

Although dental cements are used only in small quantities, they are perhaps the most important materials in clinical dentistry because of their application as luting agents to bond preformed restorations and orthodontic attachments in or on the tooth, cavity liners and bases to protect the pulp and as foundation and anchor for restorations and restorative materials.[12]

There are available cements of four basic types classified according to the matrix forming species. Classification can be seen in Table 2. Some of them contain some polymers groups.

Polymer Reinforced Zinc Oxide-Eugenol Cements

These materials have been as cementing agent for restorations, as a cavity liner and base material and as a temporary filling material. The polymer reinforced zinc oxide-eugenol cements contain 80% zinc oxide and 20% poly(methylmethacrylate), polystyrene, or polycarbonate together with accelerators in the powder. The liquid is eugenol, and accelerators such as acetic acid as well as antimicrobial agents, such as thymolor 8-hydroxyquinoline.[3,12]

EBA Cements

These materials have been used for final cementation of inlays, crowns, and bridge, for temporary fillings, and a base or lining material.

In these materials the powder is mainly 70% zinc oxide and 20% to 30% aluminium oxide or other mineral fillers. Polymeric reinforcing agents, such as poly(methymethacrylate), may also be present. The liquid consists of 50% to 60% ethoxybenzoic acid with the remainder eugenol.[3,12]

Zinc Polycarboxylate Cements

Zinc polycarboxylates are used for the cementation of cast alloy and porcelain restorations and orthodontic band, as cavity liners or base materials and as a temporary filling material.

The powder in these cements is zinc oxide with, in some cases, 1% to 5% magnesium oxide; 10% to 40% alumium oxide or other reinforcing filler. The liquid is approximately a 40% aqueous solution of polyacrylic acid copolymer with other organic acids such as itoconic acid. The zinc oxide reacts with the polyacrylic acid, forming a cross linked structure of zinc polyacrylate.[3,12]

Glass Ionomer Cements

Glass ionomer cements are another type of material that contains polymers. Glass ionomer cements are used for final cementation of cast-alloy and porcelain restorations and orthodontic bands as cavity liners or base materials and as a restorative material, especially for erosion lesions.

Table 2. Classification of dental cements

Matrix bond	Class of cement	Type
Phosphate	Zinc phosphate	Zinc phosphate
		Zinc phosphate fluoride
	Zinc silicophosphate	Zinc silicophosphate
		Zinc silicophosphatemercury
Phenolate	Zinc oxide-eugenol	Zinc oxide-eugenol
		Zinc oxide-eugenol polymer
		Zinc oxide-eugenol EBA
		Zinc oxide-eugenol alumina
	Calcium hydroxide salicylate	Calcium hydroxide salicylate
Polycarboxylate	Zinc polycarboxylate	Zinc polycarboxylate
		Zinc polycarboxylate fluoride
	Glass ionomer	Calcium aluminum polyalkenoate
		Calcium aluminum polyalkenoate zinc oxide
Polymethacrylate	Acrylic	Poly(methyl methacrylate)
	Dimethacrylate	Dimethacrylate

These materials were formulated by bringing together the siliceous material (silicate) and polyacrylate systems. The use of an acid reactive glass leads to a translucent cement that can be used as luting and filling materials.

The powder is finely ground calcium aluminium fluorosilicate glass. The liquid is a 50% aqueous solution of a polyacrylic-itoconic acid or other polycarboxylic acid copolymer that contains about 5% tartaric acid.[12]

The glass ionomer cements have several attributes over other cements with respect to their biological properties. Since they bond adhesively to tooth structure, they have potential for reducing infiltration of oral fluids at the cement-tooth interface.[10]

Polymethacrylate Based Cements

The majority of the materials in this group are poly(methacrylates) of two types materials based on methy metacrylate and materials based on aromatic dimethacrylates of the BIS-GMA type.

Acrylic Resin Cements. Acrylic resin cements are used for the cementation of restoration, facings and temporary crowns. The powder is a finely divided methyl methacrylate polymer or copolymer containing benzoyl peroxide as the initiator. The liquid is a methyl methacrylate monomer containing an amine accelerator.[12]

Dimethacrylate Cements. These cements are used for the cementation of etched-cast restorations and orthodontic bands.[12] The materials of more recent development are usually base on BIS-GMA system: they are combination of an aromatic dimethacrylate with other monomers. Such materials have been supplied as two viscous liquids, two pastes, or as powder/liquid materials. The powder is a finely divided borosilicate or silica glass containing an organic peroxide initiator. The liquid a mixture of BIS-GMA or similar aromatic dimethacrylate which is diluted with a low viscosity alkyl dimethacrylate monomer. An amine accelerator is also present.[12]

PITS AND FISSURE SEALANTS

The purpose of a pit and fissure sealant is to penetrate all cracks, pits, and fissures on the occlusal surfaces of both deciduous and permanent teeth in an attempt to seal off these suspectible areas and to provide effective protection against caries.[3] Pits and fissure sealants are polymers of the bisphenol A-Glycidyl methacrylate (BIS-GMA) type materials in which polymerization is accelerated by light or an organic amine. Sealants polimerized by visible light [420 to 450 nanometers (nm)] are one component BIS-GMA systems that require no mixing. The BIS-GMA sealants polymerized by an organic amine accelerator are supplied as two-component systems. One component contains a BIS-GMA type of monomer and a benzoyl peroxide initiator, and the second component contains a BIS-GMA type monomer with %5 organic amine accelerator. Typically, the BIS-GMA monomer is diluted with a low molecular weight dimethacrylate monomer. The two components are mixed thoroughly before being applied to the prepared teeth.[3]

REFERENCES

1. R.H. Roydhouse,1989, Introduction to Polymers, in *Dental Materials: Properties and Selection*, W.J. O'Brien, Quintessence, Chicago.
2. W.J. O'Brien, and C.L. Groh, 1989, Impression Materials, in *Dental Materials: Properties and Selection*, W.J. O'Brien, Quintessence, Chicago.
3. R.G. Craig, W.J. O'Brien, and J.M. Powers,1987, *Dental Materials: Properties and Manipulation*,4th ed., C.V.Mosby Company, St.Louis.
4. K.F. Leinfelder,1989, Composite Resins: Properties and Clinical Performance, in *Dental Materials: Properties and Selection*, W.J. O'Brien, Quintessence, Chicago.
5. K.F. Leinfelder, J. E. Lemons, 1988, *Clinical Restorative Materials and Techniques.* Lea&Febiger, Philadelphia.
6. C.L. Sockwell, H.O. Heymann,1985,Tooth-Colored Restoration in *The Art and Science of Operative Dentistry*, M. Sturdevant, R.E. Barton, C.L. Sockwell, and W.D. Strickland, 2 nd ed., C.V. Mosby Company, St Louis.
7. R.G. Craig(ed) ,1993, *Restorative Dental Materials*, 9th ed., C.V. Mosby -Year Book, St. Louis.
8. R.D. Phoenix,1996, Denture base materials, *Dent Clin North Am.*40(1):113 -120.
9. B.E. Causton,1989, Denture Base Polymers and Liners in *Dental Materials: Properties and Selection*, W.J. O'Brien, Quintessence, Chicago.
10. R.W. Phillips, 1991, *Science of Dental Materials*, 9th ed., W.B. Saunders Company, Philadelphia.
11. K.D. Rudd, 1996, Processing Complete Dentures, *Dent Clin North Am.* 40(1):121-149.
12. D.C. Smith, 1989, Dental Cements, in *Dental Materials: Properties and Selection*, W.J. O'Brien, Quintessence, Chicago.

CONTROLLED RELEASE ANTIBIOTICS FOR TREATMENT OF PERIODONTAL DISEASE

R. Darvari[1], D.L. Wise[2], V.N. Hasırcı[3], M. Boroujerdi[1], Debra J. Trantolo[4], and Joseph D. Gresser[4]

[1] Bouvé College of Pharmacy and Health Sciences
Department of Pharmaceutical Sciences
Northeastern University
Boston, MA 02115
[2] Department of Chemical Engineering
Center for Biotechnology Engineering
Northeastern University
Boston, MA 02115
[3] Biotechnology Research Unit
Department of Biological Sciences
Middle East Technical University
Ankara, Turkey
[4] Cambridge Scientific,
Inc., 195 Common Street,
Belmont, MA 02178-2909 USA

INTRODUCTION

Ciprofloxacin [1-cyclopropyl-6-fluoro-1, 4 dihydro-4-oxo-7-(1–piperazinyl)-3-quinoline carbonic acid] has a potent bactericidal activity against a broad spectrum of pathogens[1]. Poly lactide-co-glycolide (PLGA) as a biodegradable and biocompatible material has been vastly used in biomedical, pharmaceutical, and even environmental applications[2-8]. One potential application is the treatment of periodontal disease. In such an application, the small packed volume and slow flow of fluid from the cavity develops a high local concentration of the antibiotic for a prolonged time, but without the necessity for systemic dosing. Within that time the polymeric carrier will undergo hydrolytic degradation to soluble metabolizable which at the ambient pH will be largely lactate and glycolate ions.

SIGNIFICANCE

Periodontal disease involves progressive bone loss and eventual tooth loss as the supporting boney structure is eroded. To arrest the course of disease the pockets must be cleaned of severely infected tissue and tooth surfaces of calculus. Residual infection may be treated with a course of antibiotics. When thus prepared the cavity may be filled with a packing material such as powdered human bone which serves two functions. First, in conjunction with a membrane it prevents the incursion of soft tissue to the tooth surface while allowing bone regeneration. Second, the powdered bone is itself osteoinductive because of the presence of the naturally occurring bone morphogenic protein (BMP) and thus stimulates healing.

Periodontal disease which has caused serious bone erosion presents two problems: treatment of the infection and creation of an environment which allow bone regeneration. Following removal of infected soft tissue (curettage) and root planing or more advanced surgery to gain access to the tooth root surface, the cavity must be protected for incursion of soft tissue and kept free of pathogens. Exclusion of epithelium from the tooth root interface must be achieved to allow healing via regeneration of the dentogingival complex. The accepted approach, guided tissue regeneration, employs a membrane, frequently Teflon (TM), to insure the former. A packing of sterile human powdered bone can be introduced the cavity to encourage bone regeneration. Replacement of the packing material with a biocompatible, resorbable (biodegradable) polymeric matrix has several inherent advantages. First, of course, is the resorbability, true also of the powdered bone. Second, the polymer, when formulated as a powder may serve as a carrier for one or more antibiotics which will slowly diffuse into the cavity. As a powder it may be packed into the cavity to give high local concentrations of the antibiotic over extended periods; on the other hand, a small cylinder or pellet of polymer/antibiotic may be considered. Third, bone regrowth may be stimulated by incorporation of bone morphogenic protein (BMP), for which a carrier is required. Fourth, the increasing anxiety, not necessarily rational, of some patients concerning placement of human products into their bodies, other than in extreme emergencies, makes an alternative to human bone desirable.

Periodontists and orthopaedists appear to agree that patients are increasingly reluctant to receive allografts of cadaver bone due to fear of AIDS. In many instances their reluctance is not assuaged by assurances that the cadaver bone has been sterilized and confirmed to be HIV free. Thus in spite of the utility of cadaver bone, a substitute is thought by many to be necessary. Although the osteogenic potential of PLGA is questioned, it may be enhanced by incorporation of a bone morphogenic protein. Indeed, this would give the matrix osteoinductive properties; such investigations are planned for later investigations.

The polymers of choice for this application, the PLGA's, have a long history of use as resorbable suture materials and have received considerable attention as implantable vehicles for controlled release of a wide variety of therapeutic agents. Although it appears quite feasible to replace the presently used Teflon membrane with a resorbable PLGA membrane, the work discussed herein focuses only on the PLGA delivery system. Further research will focus on system optimization with respect to release rates and polymer degradation. That work will include testing in a suitable animal model. Also, the concept explored to date will be extended to include other antibiotics so that eventually the periodontal surgeon may have a variety of controlled release systems from which to choose. Further, as previously mentioned, we plan to explore incorporation of BMP into a controlled release form to induce rapid bone regeneration.

The system as visualized will replace current packing material such as hydroxyapatite and powdered human bone, providing the advantages of controlled delivery of antibiotics and resorbability. Once feasibility has been demonstrated with the antibiotics chosen for this study, it will be possible to design systems incorporating a range of antibiotics such that simple physical mixtures will enable treatment with two or more antibiotics simultaneously.

Further, as mentioned above, the matrix can also serve as a carrier of BMP to increase the rate of bone formation.

RELEASE OF ANTIOBIOTICS

Release of antibiotics from biocompatible polymeric matrices for prevention and treatment of bone infections has been of interest due to problems associated with systemic administration. Although systemic antibiotics have been administered intravenously, results are too frequently unsuccessful. High systemic levels do not necessarily achieve therapeutically effective local levels and present the risk of systemic toxicity with possible damage to kidneys or hearing[9].

Consideration of these problems led Buchholz[10] to incorporate gentamycin into a polymeric bone cement. Although reported to be successful, a problem was encountered in that large amounts of cement prevented drainage from the region of debridgement and were difficult to remove if redebridgement was necessary. This led to the development of products such as Septopal and Palacos R, which are beads of poly(methyl methacrylate), PMMA, loaded with gentamycin, and strung on surgical wire[11,12]. These could be packed into the surgical site to achieve high local concentrations without high systemic levels.

Various workers have expanded the concept to include other drugs and polymeric carriers. Kirkpatrick et al.[13] studied release of tobramycin from PMMA beads at three loadings and found, as expected, that the more highly loaded beads released more rapidly. Marks et al.[14] compared the release of three antibiotics (oxacillin, cefazolin, and gentamicin) from two commercial acrylic cements, Simplex and Palacos. Henry et al.[15] described a laboratory technique for incorporating tobramycin into the acrylic cement, Palacos. Palacos was chosen because it was said to release a higher percentage of the antibiotic over a longer interval than other bone cements.

Wahlig and Dingeldein[16] compared *in vitro* release of 12 antibiotics from five bone cements. These workers also chose the gentamycin/Palacos matrix. In most human trials, serum levels averaged 1.8 ug/ml while exudate from the wound at the implant site showed concentrations of up to 150 ug/ml.

The mechanism of release of gentamicin from PMMA was studied by Baker et al.[17]. Thin discs of PMMA cut from injection molded rods were impermeable to dissolved gentamicin and to methylene blue. *In vivo* studies using gentamicin loaded rods revealed that the more highly loaded rods released a greater proportion of drug in a given time and that these more highly loaded rods also had more defects (cracks, pores, fissures). Baker concluded that release occurred through the defects, rather than by diffusion through the polymer lattice.

The problems encountered with the PMMA preparations may be summarized as follows: 1. Lack of degradability requires eventual surgical removal. 2. Masses of cement may prevent adequate drainage. 3. Release of incorporated drug may depend on access to defects in the matrix. 4. Significant quantities of antibiotic may be trapped permanently in the matrix, if access to defects is unavailable. The advantages are also apparent: 1. High local levels of antibiotic may be achieved without high systemic levels. 2. Antibiotic release may be sustained for extended periods.

Controlled release of antibiotics from polymers has also been explored for treatment of periodontitis. Goodson et al.[18,19] have investigated the release of tetracycline hydrochloride from fibers of ethylene vinyl acetate. The fibers, 0.5 mm diameter and loaded with 25% by weight of the antibiotic, were wrapped around the tooth within the periodontal cavity. Healing observed over 60 days was significantly greater than observed for controls. Activity against six bacteria selected as periodontal pathogens was also demonstrated.

BIODEGRADABLE POLYMERS

The advantages of PMMA preparations may be retained and the disadvantages circumvented by use of a biodegradable carrier for the antibiotic. Gerhart et al.[9] reported an experimental cement comprised of poly(propylene fumarate), PPF, cross-linked with methyl methacrylate and impregnated with either vancomycin or gentamycin. (PPF, a polyester, is the site of hydrolytic degradation.) A conventional PMMA cement was used as a control. The loading of each antibiotic in polymer was 3.3% w/w. Implants were tested subcutaneously in rats. Release of antibiotic into wound fluid from the experimental and control cements was followed as a function of time following implantation. The authors state that "peak serum levels for both cements and both antibiotics were one to two orders of magnitude lower than concurrent wound levels" and that "serum antibiotic levels were unmeasurable in all cases by one week." However, in addition antibiotic levels in wound fluid were significantly higher with MMA cements than levels obtained with the control cement, PMMA. This again points to the advantage of a biodegradable carrier.

The poly(lactide-co glycolides), PLGA's, are tissue compatible and biodegradable. Much work by us and others has shown PLGA to be a versatile sustained release vehicle for a variety of drugs. Further, the osteogenic potential of PLGA was observed by Hollinger[20] in an *in vivo* study of 25 rats. PLGA implants, prepared from a 50:50 copolymer were implanted in osseous defects created in the animals' tibia. Similar defects created in their humeri received no implants and were reserved as control sites. Animals were evaluated in groups of five at 7, 14, 21, 28, and 42 days. In contrast with control sites, implant sites showed accelerated healing at 7, 14, 21, and 28 days ($p < 0.001$) with no adverse effects. Control and implant sites showed similar healing at 42 days ($p < 0.25 - 0.10$). At 42 days osseous union was complete in both experimental and control sites, with polymer resorption evident from the former.

Hollinger[20] suggests non-specific hydrolytic scission of the polymer chain, generating lactate and glycolate ions as the mechanism of resorption. Lactate is eliminated as carbon dioxide via the tricarboxylic acid cycle. Glycolate is transformed to glyoxylate via glycolate oxidase. The glyoxylate reacts via glycine transaminase to form glycine which may enter pathways of protein synthesis or be transformed sequentially to serine and pyruvate which also enters the tricarboxylic acid cycle.

The hydrolysis products of PLGA may be either acids or salts, depending on the pH at which hydrolysis occurs. Thus lactic acid and glycolic acid are formed in acid media, but lactate and glycolate ions predominate in neutral to basic media. The pH of saliva is reported to be between 6.35-6.85 (mean = 6.60); of plasma, 7.4; and of intercellular fluid, 6.1[21]. If we assume that the pH of fluids in the sulcus is that of plasma, a simple calculation gives the proportions of the salt and acid forms generated.

Both lactic and glycolic acids have a pK value of about 3.83. Thus, the ratio of acid form (lactic or glycolic acid) to salt form (lactate or glycolate) at the above pH's are:

fluid	pH	Ratio of Acid to Salt Form
Intercellular	6.10	0.00537 (0.537%)
saliva	6.60	0.00170 (0.170%)
plasma	7.40	0.00027 (0.027%)

Fluids in the sulcus have a pH essentially that of plasma; thus the acid to salt ratio will be between 0.027% or about 270 parts per million.

At the present level of development, the proposed delivery system is conceptualized as a PLGA biopolymer "powder" or "pellet" containing an antibiotic. The salient features defining the release from this delivery system are found in the characteristics of the co-polymer system and the matrix composition.

PLGA DELIVERY SYSTEMS

Polylactide, polyglycolide and the poly(lactide/glycolide) copolymers (PLGA) have been used for many years as surgical sutures because they are biodegradable, biocompatible and exhibit moderate strength in tension, compression and bending. These properties make PLGA copolymers desirable for use in many veterinary and medical applications. Over the past two decades, PLGA copolymers have been investigated extensively for the delivery and controlled release of compounds such as proteins and various pharmaceutical products.[22-25]

The physical advantages of PLGA copolymers include strength, hydrophobicity and pliability. The polymer is water insoluble but is degraded by hydrolysis to the monomers, lactic acid and glycolic acid, which are water soluble. PLGA is miscible with a wide variety of biologically active compounds. Solid formulations of polymer with a biologically active compound can therefore be prepared and designed to respond to (an) aqueous environment(s) by slowly releasing both the compound and lactic and glycolic acids.

A variety of drug/PLGA systems are currently under development as candidate systems for the long-term maintenance of therapeutic drug levels. Systems have been successfully devised for the delivery of a range of chemical classes and to achieve delivery periods that vary from days to months. Typically, these systems are fabricated in forms in which the drug is either encapsulated or incorporated into the matrix, or a combination of these two forms. The final formulation must be characterized not only in terms of its chemical composition but also with respect to physical features such as size, shape, density, and porosity which determine the drug polymer release profile.

Materials are released from PLGA-compressed matrices via a combination of diffusion and erosion. As drug particles which are solvated diffuse out of the matrix, the exposed polymer is hydrolyzed and released as monomers. New drug/matrix surface is thus exposed and the process of diffusion and erosion continues. The mechanism of release is complex. Despite extensive mathematical modeling of release profiles, no single model is able to correlate the impact of polymer formulation, drug properties, drug loading profile and dosage form on the release rates(s). Empirically, however, sufficient data are available to permit the design of PLGA systems with desired release characteristics.

PLGA polymers can be prepared in any molar ratio of lactide to glycolide acids. The proportion chosen is important in determining the *in vivo* degradation rate. Polymers prepared in a 50:50 proportion are hydrolyzed much faster than those which have a higher proportion of either monomer. For use in drug delivery systems, lactide is usually selected as the predominant species because it is more hydrophobic than glycolide.

Both the molecular weight and the molecular-weight distribution of the polymer affect the lifetime of the device. The dependence of release rate on polymer molecular weight was demonstrated in work on the sustained release of sulfadiazine from polylactide.[24] The release rate was reduced as the molecular weight of the polymer increased. It has also been shown that lowering the polymer dispersity from 2.0 to 1.4 produced a system for the release of pyrimethamine, which had a reduced initial level of release and a constant release rate for a longer period.[22]

The rate of release of a particular drug from a specific PLGA matrix can be controlled by manipulating the percentage drug content in the drug/polymer composite. This was illustrated by the release of the narcotic antagonist, naltrexone, from PLGA matrix beads.[23] Four different drug-loading percentages, ranging from 50 to 80%, were prepared in a 75/25 PLGA copolymer prepared as rods 1.6 mm in diameter. Analysis of drug release *in vitro* showed that release rates increased as loading increased. The time taken to achieve 80% release varied from 8 to 45 days as loading was decreased from 80% to 50%.

Very soluble active macromolecules, such as proteins, are released relatively rapidly from PLGA systems. Long-term delivery of proteins such as antigens for single-dose immunization systems or peptide hormones in various therapeutic settings can be achieved with PLGA matrix systems with very low drug contents.

The gross physical features of the dosage form (size and shape) are of less significance in determining the release rate. The choice is made primarily on the basis of the specific delivery requirements. Oral systems may be coated beads, powders, or cylindrical pellets. Bolus devices are often monolithic cylinders 1.9 cm in diameter. Subcutaneous or intramuscular implantation devices can be constructed as rods, beads or suspended powders. Often the ability to retrieve implanted devices after a period of use is desirable. This is accomplished most easily with rods, but is impossible with powders, and difficult for beads that are small enough to be easily implanted through a trochar.

Recent work in our laboratories has demonstrated the dependence of release rate on the state of subdivision of the polymer prior to extrusion. As an example, Figure 1 depicts the *in vitro* release of the antitubercular drug, isoniazid (INH), from matrices prepared by incorporating INH into PLGA ground and sieved to selected particle size ranges. Within the range of particle sizes examined, matrices prepared with the large polymer particles (125–180μm) release more a measure of fraction of INH released per hour.) slowly than those prepared with smaller polymer particles. Thus the linear portion of the release curve for matrices prepared with the <45μm polymer has a slope of approximately 0.009 hr⁻¹ compared to 0.0045 hr⁻¹ for matrices prepared with the 25–180μm polymer. (Slopes are a measure of fraction of INH released per hour.)

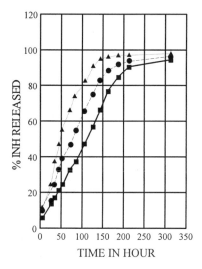

Figure 1. Percent INH Released as a Function of Time for Varying Polymer Particle Size (Δ<45μm; ◊ 45-90μm; □ 125-180μm). Note: This Figure 1 is an example of release rate as function of particle size (from our work on controlled release for TB treatment; see references 2-4.)

The prepared materials can be sterilized by gamma radiation. The Cobalt-60 process at 2.5 Mrad dose is effective and has been shown to have minimal impact on release rate[22]. Gamma ray sterilization is the most effective and practical method for sterilizing PLGA materials. Both steam and dry heat sterilization require temperatures at which the polymer would soften and flow. Gaseous sterilization (ethylene oxide, propylene oxide, formaldehyde vapor) leave residues of these compounds which are undesirable in materials to be in contact with tissues. Although degassing procedures could eliminate these residues, this adds time and expense to the manufacturing process. The effect of gamma ray sterilization of PLGA-90:10/naltrexone matrices has been examine by Wise[26]. A gamma ray dose of 2.5 Mrad, a guaranteed sterilization dose, was found to cause a reduction of about 20% in the molecular weight of the polymer. Thus a decrease in molecular weight can be compensated by using a polymer of appropriately higher molecular weight. Given an initial starting weight average molecular weight of 200,000 D, at 20% drop to 180,000 D will not substantially affect release profiles.

BONE REPAIR MATERIALS

The potential of polymeric materials has been appreciated by surgeons, dentists, and medical researchers, and promising developments have been reported. Some of the aims of work to date have been attainment of biodegradability, flexibility of formulation for tailoring at the surgical site, and achievement of ultimately acceptable esthetic results. The need for such materials is especially acute in the repair of bone defects. The overall state of development of repair materials to date, while promising, has not yet been brought to a satisfactory state. Salient characteristics expected in improved bone repair materials include: (1) controlled biodegradability; (2) primary stability; (3) compositions which range in viscosity from fluid when an injectable cement is needed, to a high viscosity formable material which may be molded to the contours required for fracture fixation; (4) acceptable working times; (5) development of adequate physical strength; and (6) potential for incorporation of sustained release antibiotics, growth factors, etc. into the repair material.

Work has been based on the use of biocompatible, biodegradable polymers synthesized from substances occurring in the Krebs cycle of metabolism. Chemical qualities of these polymers include the possibility of formulating them in a molecular weight range well-suited for preparing pastes, moldable putties, or emulsions. In addition, certain members of this class of polymers possess unsaturation, or potential for controlled cross-linking. Under proper control this cross-linking characteristic may be exploited to convert a formable mass to a rigid structure having good physical properties, but retaining the quality of biodegradability. Previous work to apply these materials to the sustained release of drugs has experimentally confirmed their biodegradability as well as revealed the preparative procedures required to provide desired repair materials[22].

It has been postulated that the success of polylactide, one of these types of polymers, as an implant material tolerated by the tissues so well is due to the fact that the breakdown product of its hydrolysis, lactic acid, is a material naturally occurring in the body. The fact that no foreign substance of disruptive chemical is produced may be of primary importance in rendering polylactide innocuous when implanted in the tissues. This reasoning led to the search for other families of polymers possessing this trait, i.e., dissolution in the tissues to produce material normally present in the body. Success was met in this search when investigation was made of polymers prepared from the substance occurring in the Krebs cycle of metabolism (otherwise termed the "citric acid cycle" or the "tricarboxylic acid cycle"). These polymers are polyesters prepared from such acids as citric, cis-aconitic, α-ketoglutaric, succinic, fumaric, malic and oxaloacetic. These acids are reacted with physiologically tolerable polyol compounds, e.g., glycerol, glycerol esters, propylene glycol, mannitol, or sorbital. Among these polymers are substances which have properties which fit well for application as medical repair or prophylactic materials. These types of materials are biocompatible and can be manipulated to yield range of materials with varying physical properties suitable for drug delivery and bone repair.

MATERIAL AND METHODS

Material

Ciprofloxacin was a generous gift from Bayer AG (Leverkusen, Germany). PLGA with a 85:15 weight ratio of lactide-to-glycolide was purchased from Boehringer Ingelheim (Germany). Solvents (glacial acetic acid acetone, and 2-propanol) were obtained from Fisher Scientific (Pittsburgh, PA). Ciprofloxacin and solvents were used without further purification. PLGA was purified by precipitation prior to use.[2-6]

Methods

PLGA Purification. The polymer was dissolved in acetone (5 g/dl) and the solution was gradually added into continuously stirred 2-propanol in a volume ratio not exceeding 100 ml of polymer solution to 50 ml of 2-propanol. The fibrous precipitate was vacuum dried (<1 mmHg) at room temperature for 24 h prior to use[2-6].

Preparation of Ciprofloxacin loaded PLGA Matrices. Purified PLGA was ground in a liquid nitrogen cooled Tekmar A-10 mill and sieved through standard Tyler mesh screens to retain the 125 180mm fractions. Ciprofloxacin powder was used without further micronization. Dry mixing method was used to blend PLGA and ciprofloxacin[2]. 0.5:1.5 and 1:1 (w/w) ratio of ciprofloxacin/micronized PLGA were mixed and blended in a 2.5″ i.d. by 2″ glass jar rotating at 65 rpm for 24 hours with no grinding aids. The final products contained 25% and 50% by weight of ciprofloxacin, respectively.

The ciprofloxacin/PLGA preparations were loaded into a 1.0 in. diameter mold equipped with a 0.043 in. (1.09 mm) diameter extrusion die. The assembled mold was mounted onto a hydraulic press equipped with a constant pressure control (Compac Model MPC 40–1, Stenhoj Co., Denmark). The pressure was controlled at 7 tons while the temperature was gradually raised with external heating tapes to the minimum at which extrusion would occur at that pressure and determined as 55°C. Cylinders obtained as the result of extrusion were cut into pieces containing 0.1 ± 0.01 g of the material for *in vitro* release studies.

In Vitro Release Study. Extruded ciprofloxacin/PLGA rods (0.1 ± 0.01 g) were immersed in 50 ml of pH 7.4 phosphate buffer saline (PBS) contained 0.1 M phosphate buffer and 0.05 M NaCl. *In vitro* release of the samples was performed in a thermostat-equipped shaking bath set at 37°C and operating at 40 cycles per minute. Each matrix was run in triplicate. Absorbance was measured at 269 nm on a Cary 1 spectrophotometer.

RESULTS AND DISCUSSION

In Vitro Release Study

Release profiles of 25% and 50% ciprofloxacin loaded extruded PLGA 85:15 are shown in Figure 2. Release kinetics were analyzed by the Rosemen-Higuchi diffusion model[27-29]. Degradation of PLGA is not considered PLGA microspheres, Tice et. al.[30] showed that PLGA-87:13 (very close to the composition of PLGA-85:15 used in this study) has a degradative half life of 4.5 months, implying a 16% degradation occurred within 35 days based on a first-order degradation. Accordingly, the Roseman-Higuchi diffusion model is applicable for the first 35 days of the release. Since after that period degradation of PLGA becomes significant, an increase in release rate is detectable. Further increase in release rate is expected after 60 days, when the bulk degradation of the polymer occurs. The last points of data from the samples (Figure 2) indicate this event. Matrices with 50% (w/w) of ciprofloxacin show higher release rates for the first 10 days in comparison with those with 25% loading of the drug. Matrices with 25% loading of ciprofloxacin show a better consistency over the increase in release rate which makes that formulation a preferred choice for sustained release applications, however, 50% loaded matrices demonstrate better reproducibility.

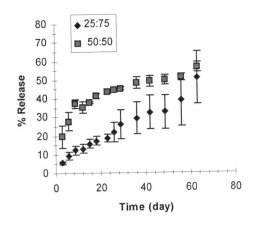

Figure 2. *In Vitro* release profiles of ciprofloxacin-loaded extruded PLGA matrices.

CONCLUSIONS

Extruding dry-mixed blends of ciprofloxacin and PLGA offers the advantage of a solvent-free formulation. The results from release profiles of biodegradable extruded PLGA matrices with two different loads of ciprofloxacin show the significant effect of loading ratio on release characteristics of these formulations. The formulated matrix has a significant effect on decreasing the release rate for periods over 60 days. In applications such as periodontal treatments or bone surgery, treatment periods of up to twenty days are preferable. Accordingly, appropriate formulations for such applications can be designed by increasing the drug portion in the matrix.

REFERENCES

1. M. LeBel, Ciprofloxacin: Chemistry Mechanism of Action, Resistance Antimicrobial Spectrum, Pharmacokinetics Clinical Trials, and Adverse Reactions, *Pharmaco-therapy*, **8**, 3-33, (1988).
2. Y. Hsu, J.D. Gresser, D.J. Trantolo, C.M. Lyons, P.R. Gangadharam, D.L. Wise, "The Effect of Polymer Foam Morphology and Density on the Kinetics of *In Vitro* Controlled Release of Isoniazid from Compressed Foam Matrices", *J. Biomed. Mat. Res.*, in press.

3. Y.-Y. Hsu, J.D. Gresser, D.J. Trantolo, C.M. Lyons, P.R.J. Gangadharam, and D.L. Wise, "Low Density Poly(DL-Lactide-co-Glycolide) Foams for Prolonged Release of Isoniazid," *J. Contr. Rel.*, **40**, 293-302, (1996).

4. Y.-Y. Hsu, J.D. Gresser, R.R. Stewart, D.J. Trantolo, C.M. Lyons, G.A. Simons, P.R.J. Gangadharam and D.L. Wise, "Mechanisms of Isoniazid Release from Poly(D,L-lactide-co-glycolide) Matrices Prepared by Dry Mixing and Low Density Polymeric Foam Methods", *J. Pharm. Sci.*, **85**, 706-713, (1996).

5. D.J. Trantolo, J.D. Gresser, L. Yang, D.L. Wise, J.F. Smith and P.J. Giannasca, "Delivery of Vaccines by Biodegradable Polymeric Microparticles with Bioadhesion Properties", *Proceedings of the 5th World Congress Chem. Eng.*, (1996).

6. Y.-Y. Hsu, J.D. Gresser, D.J. Trantolo, C.M. Lyons, and D.L. Wise, "*In Vitro* Controlled Release of Isoniazid from Poly(lactide-co-glycolide) Matrices," *J. Cont. Rel.*, **31**, 223, (1994).

7. R. Darvari, V.N. Hasirci, "Pesticide and Model Drug Release from Carboxymethyl Cellulose Microspheres" (in press, *J. Microencapsulation*).

8. I. Gursel, V.N. Hasirci, "Properties and Drug Release Behaviour of PHB and Various P(HB-HV) Copolymer Microcapsulates", *J. Microencapsulation*, **12** (2), 185-193, (1995).

9. T.N. Gerhart, R.D. Roux, G. Horowitz, R.L. Miller, P. Hanff, and W.C. Hayes, "Antibiotic Release from an Experimental Biodegradable Bone Cement", *Journal of Orthopaedic Research*, **6**, 585, (1988).

10. H. Buchholz and E. Engelbrecht, "Uber die Depotwirkung einiger Antibiotikabei Vermischungmit dem Kunstharz Palacos", *Chirurg.*, **41**, 511, (1970).

11. J.A. von Fraunhofer, H.C. Polk, and D. Seligson, "Leaching of Tobramycin from PMMA Bone Cement Beads", *Journal of Biomedical Materials Research*, **19**, 751, (1985).

12. J.A. Goodell, A.B. Flick, J.C. Herbert, J.G. Howe, "Preparation and Release Characteristics of Tobramycin-Impregnated Polymethylmeth-acrylate Beads", *Am. J. Hosp. Pharm.*, **43**, 1454, (1986).

13. D.K. Kirkpatrick, L.S. Trahtenberg, P.D. Mangino, J.A. von Fraunhofer, and D. Seligson, "*In Vitro* Characteristics of Tobramycin-PMMA Beads: Compressive Strength and Leaching", *Orthopedics*, **8**, 1130, (1985).

14. K.E. Marks, C.L. Nelson, and E.P. Lautenschlahger, "Antibiotic-Impregnated Acrylic Bone Cement", *Journal of Bone and Joint Surgery*, **58**, 358, (1976).

15. S.L. Henry, G.J. Popham, P. Mangino, and D. Seligson, "Antibiotic-Impregnated Beads: A Production Technique", *Contemporary Orthopaedics*, **19**, 221, (1989).

16. H. Wahlig and E. Digeldein, "Antibiotics and Bone Cements", *Acta Orthop. Scand.*, **51**, 49, (1980).

17. A.S. Baker and L.W. Greenham, "Release of Gentamicin from Acrylic Bone Cement", *J. of Bone and Joint Surgery*, **70-A**, 1551, (1988).

18. J.M. Goodson, et al., "Multicenter Evaluation of Tetracycline Fiber Therapy: I. Experimental Design, Methods, and Baseline Data", *J. Periodontal Research*, **26**, 361, (1991).

19. J.M. Goodson, et al., "Multicenter Evaluation of Tetracycline Fiber Therapy: II. Clinical Response", *J. Periodontal Research*, **26**, 371, (1991).

20. J.O. Hollinger, "Preliminary Report on the Osteogenic Potential of a Biodegradable Copolymer of Polylactide (PLA) and Polyglycolide (PGA)", *J. Biomedical Materials Research*, 17, 71-82, (1983).

21. Biochemistry, Albert L. Lehninger, Worth Publishing Co., N. Y. p. 46, (1970).

22. D.L.Wise,(ed.),Biopolymeric Controlled Release Systems, **I & II**, CRC Press, Boca Raton, FL.

23. D.L. Wise, R.H. Reuning, S.H.T. Liao, A.E. Staubus, S.B. Ashcraft, D.A. Downs, S.E. Harrigan and J.N. Wiley, "Pharmaco-kinetic Quantitation of Naltrexone Controlled Release

from a Copolymer Delivery System", *J. Pharmacokinetics Biopharm.*, **2**, (1983).

24. D.L. Wise, J.D. Gresser and G.J. McCormick, "Sustained Release of a Dual Antimalarial System", *J. Pharm. Pharmacol.*, **31**, 201-204, (1979).

25. D.L. Wise, "Biopolymer System Design for Sustained Release of Biologically Active Agents", Chapter 1 in D. L. Wise, editor, *Biopolymeric Controlled Release Systems*, **I**, CRC Press, Inc., (1984).

26. D.L. Wise, "Development of Drug Delivery Systems for Use in Treatment of Narcotic Addiction--A Culmination", Chapter 8 in Biopolymeric Controlled Release Systems, 1, D.L. Wise, Ed., CRC Press, Boca Raton, FL, (1984).

27. T. Higuchi, "Physical Chemical Analysis of Percutaneous Absorption Processes from Creams and Ointments", *J. Soc. Cosmetic Chemists*, **11**, 85-97, (1960).

28. T. Higuchi, "Rate of Release of Medicaments from Ointment Bases Containing Drugs in Suspension", *J. Pharm. Sci.*, **52**, 1145-1149, (1963).

29. T.J. Roseman and W. I. Higuchi, "Release of Medroxyprogresterone from a Silicone Polymer", *J. Pharm. Sci.*, **59,** 353-357, (1970).

30. T.R. Tice, D.H. Lewis, R.L. Dunn, W.E. Meyers, R.A. Casper and D.R. Cowsar, Biodegradation of Microspheres and Biomedical Devices Prepared with Reabsorbable Polyesters, *Proc. Int. Symp. Control. Rel. Bioact. Mater.*, 9-210, (1982).

MICROBIAL POLYHYDROXYALKANOATES AS BIODEGRADABLE DRUG RELEASE MATERIALS

Vasıf Hasırcı[1], İhsan Gürsel[1], Füsün Türesin[1], Gökçe Yiğitel[1], Feza Korkusuz[2], and Gürdal Alaeddinoğlu[1]

[1]Department of Biological Sciences, Biotechnology Research Unit
[2]Medical Center
Middle East Technical University Ankara 06531, Turkey

INTRODUCTION

The advantages of the use of controlled release systems for the delivery of bioactive species are numerous. To name a few of these, increased effectiveness of the delivery, achievement of high local concentrations of drug, increased stability of especially labile drugs, the resultant cost effectiveness of the therapy can be stated. As a result of the realisation of these advantages and the substantial increase in the number and kind of materials from which these can be constructed, the applications of Controlled Drug Delivery Systems became the hottest topic of Biomaterials and Pharmaceutical Technology fields. The delivery system almost always consists of a carrier and the bioactive species the relation of which with respect to each other determines the release behaviour. The earlier drug release systems employed synthetic polymers as the carrier and upon depletion of their drug contents there remained an empty shell or a casing which had to be disposed of, a problem if the release system employed was implanted. Development of biodegradable polymers of biological and synthetic origin led to a new wave of research. both due to new mechanisms of release which now became possible and also due to avoidance of the retrieval problem.

The use of biodegradable implantable drug release systems, in general, may lead to an overall better quality of life due to:
•fewer surgeries (degradability making retrieval surgeries unnecessary)
•better recovery (by locally providing controlled and desired dose of antibiotics)
•more efficient use of hospital facilities and personnel
•lower probability of infection
•production of less medical waste, and
•use of lower quantities of drugs

When the drug release system is constructed of a biological origined material, an added benefit would be conservation of non-renewable sources (eg. petroleum products such as poly(lactic acid), PLA; poly(lactic-co-glycolic acid), PLGA; poly(methyl metahcrylate), PMMA; poly(vinyl alcohol), PVA; poly(ethylene), PE; etc.). The biological

Biomedical Science and Technology
Edited by Hıncal and Kas, Plenum Press, New York, 1998

macromolecules that have been in biomedical use are, in addition to the well known natural rubber, mainly proteins (ie. collagen and gelatin) and polysaccharides (alginates, chondroitin and hyaluronic acids, cellulosics like carboxymethylcellulose and hydroxypropylcellulose). These are generally used after modification and reconstitution and as such do not lead to a large variety of materials. Another emerging category with immense potential is the microbial polymers.

The observation of a large number microorganisms , including *Bacillus megaterium*, soil bacteria, estuarine microflora, blue-green algae , producing lipophilic inclusion bodies was made in the 1920's, so they were actually available before the tailor-made industrial polymers appeared in the 1930's [1]. The realisation that polyesters make up the bulk of these bodies unveiled a great potential for use of these materials as alternatives to polymers.of petrochemical origin. These were identified to be mainly poly(3-hydroxyalkanoates) among which optically active polymers of D(-)-3-hydroxybutyric acid , P(3HB), (see chemical formula below) constituted the major portion:

$$
\begin{array}{ccccc}
CH_3 & O & CH_3 & O & O \\
| & || & | & || & || \\
HO\text{-}CH\text{-}CH_2\text{-}C\text{--}[\text{-}O\text{-}CH\text{-}CH_2\text{-}C\text{--}]_x\text{-}O\text{-}CH_2\text{-}C\text{-}OH
\end{array}
$$

Replacement of CH_3 with ethyl or pentyl groups yields poly(3-hydroxyvalerate),P3HB, and poly(3-hydroxyoctanoate),P3HO , respectively. A major advantage of these materials over the synthetic ones is the ability of the bacteria to be able to synthesize a large variety of polymers based on the the feed medium composition and the species producing the polyester. Thus, introduction of 3-hydroxypropionate, 4-hydroxybutyrate, 3-hydroxyvalerate etc. into the growth medium led to polyesters with these as the monomeric units alongside the main unit , 3-hydroxybutyrate[2]. The relative proportion of these units could be varied by their composition in the growth medium, too. Thus, by the discovery of these microbial polyesters a very versatile biopolymer production capability was unleashed.

Polyhydroxyalkanoates are semicrystalline (60-80% crytallinity[3]) and this is very important from the physical, thermodynamic, optical and mechanical and degradability point of view. It also substantially affects drug compatibility, drug diffusion and most important of all, degradation which all influence drug release properties. Present view is that poly(3-hydroxybutrate-co-3-hydroxyvalerate), PHBV, copolymers have similar degrees of crystallinity regardless of hydroxyvalerate content.

An advantage (and in some cases a disadvantage) of biopolymers is their degradability in the biological medium. It is an advantage especially when the removal of the biomedical device or system is needed upon completion of the function. The removal process, which will be an additional risk to the patient, can be avoided when the selected material degrades after fulfilment of the service. This property could especially be of immense value in the construction of novel controlled drug release systems (intended for various routes of implantation like periodontal and intramuscular) and in hard tissue fracture fixation[4].

An area where controlled release could be of substantial use is the treatment of osteomyelitis. Total loss of manpower due to osteomyelitis and fractures is enormously high. Eighty five percent of the people who are involved in traffic accidents have one or more fracture in their long bones and at least 50% of them undergo surgery related to that fracture. The risk of infection in open fractures is also high and require special attention to prevent or to treat osteomyelitis. When metal implants are used as internal fixation devices, the risk of infection increases due to the adherence of bacteria to the metal and following recovery a second operation is needed to remove the internal fixation device.

Determination of the number of implants used and their cost has never been easy and the values obtained have to be used with caution. The most reliable figures obtained from the State Statistics Institute of Turkey for 1994 indicate that 2 840 kg of medical plates and pins were imported for a total of $ 1.1 million . Thus, it is apparent from these figures that a large number of people undergo surgery requiring these plates and pins and face the risk of infection. Systemic application of antibiotics to treat these infections have never been completely succesful. Use of synthetic polymeric beads of PMMA as sustained release formulations in the treatment of osteomyelitis necessitates removal after depletion of the drug content due to the stability of the glassy acrylic polymer. A bioresorbable antibiotic loaded implant or better yet, a biodegradable, drug loaded fracture fixation rod that diminishes the need for a second surgery while removing the risk of infection, therefore, will positively influence the outcome of osteomyelitis treatment. Polyhydroxyalkanoates via erosion in the biological medium have the potential for use under such circumstances.

The ultimate aim of the present study was to introduce these novel materials to the biomedical field and to be able present them as non-degradable drug release systems for the therapy of osteomyelitis.

MATERIALS AND METHODS

Biopolymer Production

Biopolymer production is one of the parameters that is very influential on the successful use of the resultant product. Microbial production of a polyester, poly(3-hydroxybutyrate-co-valerate), with varying degrees of valerate contents leads to products with varying   physical and mechanical properties . Thus, choice of growth media, its composition and the microorganism are very influential on the resultant product. Table 1 shows selected growth media, the resultant polyesters and some of their properties.

Polymer Purification

Polymer purification is important both because the purity of the final polymer will be influential on the responses (i.e.immunological) elicited by this material and also because the purification method will determine the properties of the product (eg. use of hypochloride for breakdown of the granular capsule and thus purification decreases molecular weight).

Preparation of Sulperazon Loaded Rods

PHBV (7, 14 and 22% HV mol content) were dissolved in chloroform and then sulperazon granules were added . The resultant mixture was turned into a paste and molded to obtain rods containing PHBV: Sulperazon loadings of 5:1 and 1:1.

In Vitro Sulperazon Release

PHBV rods in predetermined dimensions were placed in dialysis bags. The bags were placed in isotonic phosphate buffer (0.1M, pH7.4), PBS, some of which was also added into the dialysis bag and stirred continuously at ambient temperature.
Antibiotic release was followed by measuring the UV absorbances at 211.5 nm.

In Vivo Experiments

A metal implant was inserted to the right hind tibia of three month old albino rabbits (New Zealand) and *Staphylococcus aureus* (obtained from chronic osteomyelitis patients) was inoculated to the same area with the implant. Three weeks later the inoculation area was evaluated (with radiological, histological and microbiological methods) for the presence of infection. PHBV implants were introduced to the infection site. After 15 and 30 days, implantation area were examined and the effectiveness of the implants were assessed by the clearance of the bacteria.

RESULTS AND DISCUSSIONS

Biopolymer Preparation

Biopolymers were prepared by varying the composition of the growth media and the microorganisms used to produce them. It was thus possible to modify the composition of the polymer chains to include 3-hydroxyvalerate and 4-hydroxybutyrate along with 3-hydroxybutyrate. This obviously has a significant influence on the crystallinity and, therefore, on the mechanical properties of the product.

Purification method also had an influence. If instead of exhaustive Soxhlet extraction of the polymer in chloroform, hypochloride was used, a decreased molecular weight was observed. This , as a result, did influence the solubility and the mechanical properties of the material.

Table 1. The microorganisms which were used to produce polyhydroxyalkanoates and the properties of the resultant polymers

Microorganism	Growth Medium	DSC/IR/NMR Conclusion
A.latus	S/4HB	PHBV9
A.eutrophus	F	Pure P3HB
A.latus	S/4HB Hypo. Treated	Hypochloride had almost no effect
Alcaligenes	Commercial	PHBV14
A.eutrophus	F/VL	PHBV11
A.eutrophus	F/4HB	10% 4HB
A.eutrophus	F/BrPPA	

BrPPA: bromopropionic acid; VL: valerolactone; HB: hydroxybutyric acid; A.eutrophus: Alcaligenes eutrophus, A.latus: Alcaligenes latus, S:Sucrose, F: Fructose

In Vitro Antibiotic Release

Antibiotic loaded rod preparation employs a paste and solvent evaporation leaves behind a polymer-drug composite. Such a structure obviously does not yield a rate controlling barrier to lead to a Zero-Order release. It was observed that release was not zero order, as it can be deduced from the structure of the implant (Figure 1). About 70% of the antibiotic was released in 54 days. This, however, reveals a significantly slow rate of release which for the purpose of the present application could suffice. Use of a more crystalline polymer or a higher polymer:drug ratio reduced this rate and the initial burst further .

In Vivo Experiments

It was observed that 15 days after implantation the antibiotic loaded rods of PHBV22 type led to the eradication of infection. With the drug-free PHBV rods S.aureus presence was unmistakable. Similar observations were made with the animals which were sacrificed after 30 days revealing the effectiveness of the therapy.

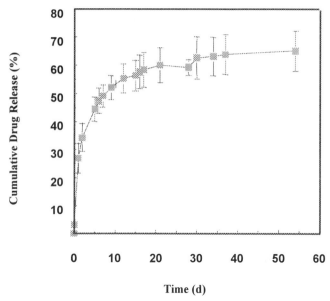

Time (d)

Figure 1. Sulperazon release from PHBV 22 Rods into PBS at RT.

Further studies, including the resorption time of the implant, *in vivo* degradation and amount of drug released, surface changes and its biocompatibility are being studied.

ACKNOWLEDGEMENTS

This manuscript originates from the NATO SFS project TU-POLYESTERS that is being carried out mainly at METU in collaboration with colleagues at various Universities and Hospitals in Ankara. The contributions by Prof. E. Bayramlı and F.Yagmurlu, M.D are especially acknowledged .

REFERENCES

1. M. Lemoigne, Etudes sur l'autolyse microbienne acidification par formation d'ucide β-oxybutyrique, *Ann. Inst. Pasteur (Paris)*,3 9:144-173 (1925).

2. C.W. Pouton and S. Akhtar, Biosynthetic polyhydoxyalkanoates and their potential in drug delivery, *Adv. Drug Delivery Rev.*, 18:133-162 (1996).

3. S. Bloembergen, D.A. Holden, G.K. Hamer, T.L. Bluhm and R.H. Marchessault, Studies of composition and crystallinity of bacterial poly(β-hydroxybutyrate-co-β-hydroxyvalerate), *Macromolecules*, 19:2865-2871 (1986).

4. I. Gürsel and V. Hasirci, Properties and Drug Release Behaviour of PHB and Various P(HB-HV) Copolymer Microcapsules, *J.Microencapsulation*, 12(2):185-193 (1995).

VANCOMYCIN-LOADED CALCIUM SULFATE FOR THE TREATMENT OF OSTEOMYELITIS - CONTROLLED RELEASE BY A POLY(LACTIDE-CO- GLYCOLIDE) POLYMER

Marie-Ange Benoit,[1] Benoît Mousset,[1] Richard Bouillet,[2] Christian Delloye,[3] and Jean Gillard[1]

[1] Université Catholique de Louvain
   Ecole de Pharmacie, Laboratoire de Pharmacie Galénique
   Industrielle & Officinale - Av. E. Mounier
   73.20 - 1200 Bruxelles, Belgium
[2] Clinique Europe - St Michel - Square Marie-Louise
   59 - 1040 Bruxelles, Belgium
[3] Cliniques Universitaires St-Luc
   Service d'Orthopédie
   Av. Hippocrate, 10 - 1200 Bruxelles, Belgium

INTRODUCTION

Despite the availability of more recent antimicrobial agents, the treatment of bone diseases like osteomyelitis, caused by microbial infection remains an important orthopaedic problem. The characteristics of bone make chronic osteomyelitis refractory. Hard walls surround the soft tissues of bone and inflammation of the contained tissues causes circulatory disturbances which can readily lead to necrosis of various parts of the bone. These anatomical features provide an environment suited for the localization and colonization by bacteria[1]. Then, the surgical removal of necrotic tissues and the antibiotic administration are the primary methods of treatment of chronic osteomyelitis[2]. Necrotic bone provides an appropriate surface for the development of a biofilm[3] and the causative bacteria produce large amount of extracellular fibrous glycocalyx materials[4]. Therefore, the antibiotic concentration must be many times higher than the MIC (Minimum Inhibitory Concentration) to eradicate bacteria encased in this biofilm[5]. This may be one reason why the disease is uneasy to treat even though the penetration of some antibiotics into infected bone is higher than into normal bone[6].

Since it is impractical to deliver antibiotics to the target site at an effective concentration by the systemic route, local administration is considered to be essential. Among these techniques, mechanical methods include isolated perfusion of an extremity[7], suction irrigation[8] and the use of an implantable drug pump[9].

Another approach is to mix antibiotics with a carrier material which can locally release the drug. Within this framework, non-biodegradable polymethylmethacrylate (PMMA) bone cement or beads have been extensively used for the treatment as well as prophylaxis of bone infection[10]. The necessary surgical removal of these beads after implantation, incomplete and poorly controlled *in vitro* release of the antibiotics are the major disadvantages of this localized delivery system. Moreover, the polymer is known to be unsuitable for the incorporation of some antibiotics including chloramphenicol and tetracycline[11].

Biomedical Science and Technology
Edited by Hıncal and Kas, Plenum Press, New York, 1998

Because biodegradable carriers obviate the need for surgical removal, they offer a potential advantage compared with the others[12,13,14,15,16].

Bone grafts associated or not to ceramics, polymeric systems and plaster of Paris are biodegradable and moldable materials which have also been used as carriers[17,18].

The human donor material is very interesting because it can recover the defective blood irrigation and so, promotes the action of the systemic administered antibiotic. Nevertheless, this allogencic graft is not always available in sufficient quantity, is susceptible to immunogenic responses of the host and may carry the risk of transmitted viral agents. There is a need, therefore, for stable immunologically privileged bone repair materials.

The PPF-MMA (poly(propylene fumarate)-methylmethacrylate) and poly(esters) appear to be good carriers of antibiotics for the treatment of osteomyelitis[12,14,15]. They can be loaded with antibiotics which are nearly released as they degrade (in contrast to the limited amount that can diffuse out the PMMA)[13,19,20,21].

Plaster of Paris, dried calcium sulfate hemihydrate, was selected as antibiotic-carrier due to its cheapness, facilities of use and sterilization[26], biocompatibility and bioresorbability. This carrier which was already used for the filling of bone cavities for more than one century[22] may be associated with various antibiotics[23,24,25,26,27,28] and osteoinductive agents[18] to produce reproducible delivery systems. Its successfull use for the treatment of chronic and acute human osteomyelitis was already related[29,30,31,32].

Among the numerous available antimicrobial agents, vancomycin was chosen because it possesses an extremely broad antibacterial spectrum. It is highly effective against Gram-positive skin flora including *Staphylococcus aureus*, S. *epidermis* (and other coagulase-negative staphylococci), streptococci, and corynebacterium, which constitue about 75 % of the bacteria implicated in deep prosthetic infections[33]. Thanks to its thermostability at 37°C, this antibiotic which was already associated to PMMA cements is suitable for being incorporated into implantable carriers[25,33]. Moreover, the glycopeptides such as teicoplanin and vancomycin were not inactivated by the heat of polymerization of plaster of Paris, retained their antimicrobial properties after incorporation into this carrier for more than six months and were readily eluted *in vitro* into the surrounding medium[23,25,34].

Among the artificial polymers which have been recognized as biodegradable so far, aliphatic polyesters of the poly (α-hydroxy acid)-type which are derived from metabolites such as lactic acid enantiomers (LA) and glycolic acid (GA), have been investigated extensively for the treatment of bone diseases and used currently as suture materials, implants and drug delivery systems[15,34,35,36,37,38]. Thanks to its adjustable biodegradation, this type of polymer was selected to try to reduce the initial burst effect of drug release from plaster of Paris implants and to extend the therapeutic release from one week to more than one month.

Finally, the purpose of the present study was to investigate *in vitro* and *in vivo* the release characteristics of vancomycin hydrochloride from plaster of Paris implants coated or not with a biodegradable polymer.

MATERIALS AND METHODS

Materials

Plaster of Paris, a dried calcium sulfate hemihydrate ($CaSO_4.1/2 H_2O$) was purchased from Merck, Darmstadt, Germany. Vancomycin chloride (Vancocin® CP) was supplied by Eli Lilly, Brussels, Belgium. $PLA_{45}GA_{10}$ (95,000 M.W.) composed by 10% (w/w) polyglycolic acid and 90% (w/w) racemic poly(D,L-lactic acid) was purchased from Phusis, Saint-Ismier, France.

Preparation of Implants

Plaster of Paris was previously sterilized by heating to 121°C in an oven for 24 hours. Implants were manufactured in a vertical laminar flow bench as previously described[26]. Briefly, sterile plaster of Paris (10 g), antibiotic (600 mg) and ultra pure water (6 ml) were mixed together and placed into sterile moulds. After drying at room temperature, the implants were stored in dry and sterile conditions until assayed.

Coating of Implants

The coating was performed on a vertical laminar flow bench by soaking implants in a 5% (w/v) $PLA_{45}GA_{10}$ solution of acetone. Four groups of implants were prepared without or with one, two or six successive soakings, giving one group devoided of coating and three others groups with an increasing homogeneous depth of coating (69, 94 and 162 μm) measured on histological slices by a histometric method.

In Vitro Release

As described previously[25], five antibiotic-loaded implants were weighed, placed in 2 ml of Phosphate Buffer Saline (PBS) (0.01 M phosphate buffered saline - NaCl 0.138 M - KCl 0.0027 M) at pH 7.4 and stored at 37°C. The buffer solution was removed daily and frozen at - 20°C until assayed. The implants were washed once with the same fresh buffer, then placed in 2 ml of PBS (pH 7.4) and incubated again.These experiments were conducted twice for each type of implant and lasted three weeks.

In Vivo Release

Three groups of eight-months old skeletally mature New Zealand white rabbits (3.5 kg ± 0.2 (n=35)) were used for the *in vivo* experiments. The plaster of Paris implant loaded with antibiotic was inserted into the femoral condyle. The animals were killed after 1 hour and 1, 3, 4, 7, 10 14 24 or 28 days of implantation. The femur was then harvested and the implant removed, weighed, ground, incubated under gentle agitation at room temperature in 5 volumes of PBS pH 7.4 for two hours. The antibiotic concentration was measured into supernatant fluids obtained by centrifugation (3,000 x g, 10 min.).

After removal of soft tissues and blood from the surface of the bone and drying, a slice of 7 mm in depth was cut below the cavity of the implant and the antibiotic concentration was measured separately in the cancellous and cortical bone specimens. After washing of each bone samples to remove blood, they were pulverised, weighed and homogenized with 5 volumes of PBS pH 7.4. After an incubation of 48 hours at room temperature under gentle agitation, the supernatant fluids were collected by centrifugation (3,000 x g , 10 min.) for antibiotic evaluation.

Blood samples were withdrawn during 24 hours after implantation to measure the serum concentration of the antibiotic.

Vancomycin Assay

Vancomycin was assayed by a FPIA (Fluorescence Polarization Immuno Assay) using the $TDx^{®}$ Abbott apparatus. The lower detected concentration is around 0.5 to 1 μg/ml. The antimicrobial activity of the elution fluids was also measured by an agar well plate method according to the method of Grove and Randall[39] using *Staphylococcus aureus* (ATCC 25923).

RESULTS

In Vitro Release Characteristics

Four types of plaster of Paris implants loaded with vancomycin (60 mg/g) were prepared in aseptic conditions. Fig. 2 shows the cumulative release of vancomycin from a control group devoided of coating and from three others respectively coated with one (69 μm), two (94 μm) and six layers (162 μm) of $PLA_{45}GA_{10}$.

Whatever the type of implants, vancomycin retained its antimicrobial effect after incorporation into plaster of Paris and was subsequently released to the surrounding medium at concentrations sufficient to inhibit bacterial growth during twenty days (daily release always superior to 50 μg/g of plaster). Moreover, the antibiotic release was inversely proportional to the coating depth : faster without coating and slower with six layers of polymer.

The release of vancomycin from the uncoated implants used as control group, was characterized by a burst effect with 70% (w/w) of the total recovered amount of drug

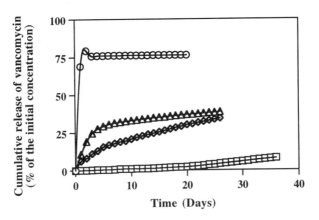

Figure 1. *In vitro* release profiles of vancomycin from plaster of Paris implants without coating (O) or coated with one (69 μm) (△), two (94 μm) (◇) or six layers (162 μm) (□) of PLA$_{45}$GA$_{10}$, in Phosphate Buffer Saline pH 7.4 at 37°C.

detected at the first day of experiment.The cumulative percentage of antibiotic release from these uncoated implants did not reach 100%. This phenomenon was due to the adsorption of vancomycin to the carrier corresponding to the amount of drug (± 20% of the initial concentration) which could not be extracted from beads in our experimental conditions.

The antibiotic loaded into implants coated with one layer (69 μm) of poly(α-hydroxy acid) polymer, was detected in the incubation medium during twenty-six days with 25% (w/w) of the initial concentration of drug released during the first two days. After three weeks of incubation, 40% (w/w) of the initial concentration of vancomycin remained into implants.

The plaster of Paris pellets coated with two layers (94 μm) of biodegradable PLA$_{45}$GA$_{10}$ presented a release profile relatively similar to the profile from the implants coated with only one layer of polymer : occurring during twenty-six days but characterized by an initial release of 8% (w/w). At the end of the study, on day twenty-six, only 43.5% (w/w) of the total amount of antibiotic incorporated into the pellets, were diffused in the incubation buffer.

A daily release of around 60 μg of vancomycin/g of implant occurred during the first two weeks when the implants coated with six layers (162 μm) were incubated in the aqueous medium. During the third week and for two weeks again the release profile became practically linear with a release rate in the range of 215 μg of vancomycin/day/g of plaster. These dose rates seemed very interesting for local antibiotherapy because the local concentration of antibiotic could reach a level higher than MIC for a lot of pathogenic germs without incurring of dangerous systemic levels. Finally, the cumulative amounts of vancomycin released after thirty-seven days were only around 10% (w/w) of the initial concentration.

In Vivo Release Characteristics

The *in vivo* study was carried out on plaster of Paris implants loaded with vancomycin (60 mg/g of plaster) devoided of coating or coated with two (94 μm) and six layers (162 μm) of PLA$_{45}$GA$_{10}$.

The release kinetic of the antibiotic has been established from its *residual concentration* remaining into the pellet after increasing times of implantation into the femoral condyle of rabbits (Figure 2).

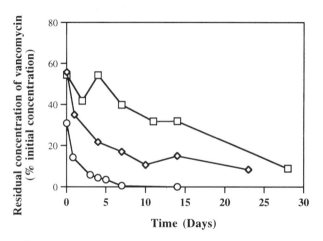

Figure 2. *In vivo* release profiles of vancomycin from plaster of Paris implants without coating (O) or coated with two (94 μm) ($\Diamond$) or six layers (162 μm) ($\square$) of $PLA_{45}GA_{10}$, inserted into the femoral condyle from rabbits.

One week after implantation, vancomycin was totally disappeared from the uncoated pellets with a high release (37% (w/w) of the initial concentration) occurring during the first hour of implantation. Nevertheless, the complete resorption of the implant monitored by radiography and histological methods requires three weeks. The presence of a biodegradable polymer coating clearly decreased the drug release at the beginning of the experiment allowing to extend the antimicrobial activity of the pellets from one week to more than four weeks. In coating with two layers (94 μm) of $PLA_{45}GA_{10}$ therapeutic concentrations were measured during at least three weeks. At the end of the experiment, 8% (w/w) of the initial concentration of vancomycin were always detected into the implanted pellets. By increasing the coating depth (162 μm), the release occurred for more than four weeks with 40% (w/w) of the initial concentration remaining in the pellets after one week of implantation. Whatever the coating thickness, the implants entirely surrounded by fibrous connective tissue were morphologically unchanged after eight weeks of implantation (data not shown).

The results of the vancomycin concentrations measured into *bone marrow and cortical bone* samples from rabbits treated with plaster of Paris pellets coated or not with $PLA_{45}GA_{10}$ (two or six layers) are summarized in table 1. They are presented in terms of antibiotic concentration upper or lower to MIC. The MIC of vancomycin against *Staphylococcus aureus* one of the major causative bacteria of osteomyelitis, is around 3 μg/g.

Whatever the type of plaster implant coated or not with $PLA_{45}GA_{10}$, the penetration of vancomycin into the bone tissues located near the implant occurred. The concentration exceeded the MIC for at least four days after implantation of uncoated vancomycin-loaded plaster pellets reaching respectively 40 and 20 μg/g into the bone marrow and cortical bone samples after four days of implantation. The coating of implants allowed to sustain a therapeutic release for at least three weeks in the two types of bone tissues. Nevertheless, the bone concentrations were always slightly lower. So, the maximal concentration of vancomycin detected into the cortical bone and bone marrow slices on day ten after of implantation of plaster pellets coated with two layers (94 μm) were respectively twice and three times lower. The antibiotic concentrations of the two types of bone tissues treated with the pellets coated with six layers (162 μm) of polymer were always around the MIC value ($\pm$ 5 μg/g) excepted at the seventh day of experimentation where 40 μg of vancomycin/g of cortical bone were detected.

The *serum concentration* of the glycopeptidic antibiotic reached a peak (0.6 μg/ml) three hours after implantation of an uncoated implant but remained undetectable throughout the study. Vancomycin was always undetectable in the serum when coated plaster of Paris implants were placed into the femoral bone of rabbits.

Table1. Detection of vancomycin into the bone marrow and cortical bone samples

Time after implantation	Uncoated implant		$PLA_{45}GA_{10}$ coating Two layers (94 μm)		$PLA_{45}GA_{10}$ coating Six layers (162 μm)	
	Bone marrow	Cortical bone	Bone marrow	Cortical bone	Bone marrow	Cortical bone
1 hour	+	+	+	+	±	+
1 day	+	+	+	+	+	±
4 days	+	+	+	+	-	-
7 days	-	-	±	±	+	±
10 days	ND	ND	+	+	+	±
14 days	-	-	+	+	±	±
21 days	ND	ND	+	+	ND	ND
28 days	ND	ND	ND	ND	±	+

+ : concentration upper to MIC (> 4 μg/g); ± : Concentration near to MIC (from 1.5 to 4 μg/g);
- : undetectable; ND : non done

DISCUSSION

This study was intended to evaluate in a controlled setting the possibility of using plaster of Paris as a vehicle for the local administration of antibiotics. It also concerns the controlled release by coating the implants with a biodegradable polymer composed by 10 % (w/w) polyglycolic acid and 90 % (w/w) racemic poly (D,L - lactic acid).

From the *in vitro* study, it turns out that the elution of vancomycin was controlled by coating plaster of Paris implants with a poly(lactide-*co*-glycolide) polymer. This biodegradable polyester was selected to embed the carrier because it does not hinder the normal osteogenesis[15,16].

The vancomycin release was closely dependent on the coating depth : therapeutic concentrations were measured during twenty-five days with a burst effect occurring during the first five days when one layer of $PLA_{45}GA_{10}$ (69 μm) coated the implants; the presence of two layers of $PLA_{45}GA_{10}$ (94 μm) allowed to decrease the initial release and to increase the daily amount of vancomycin released; according to the impermeability of the $PLA_{45}GA_{10}$ to vancomycin, resulting of the hydrophobicity of the polymer, the six layers embedded implants were characterized by a sustained release of antibiotic throughout the experiment.

These preliminary observations suggest that antibiotic is released through an interconnecting series of voids and cracks in the polymer which could be seen on histological slices (data not shown). These ones could occur through local stresses and shrinkage at the time of vaporization of the solvent polymer. So, the greater thickness of coating made the diffusion of antibiotic through defects in the matrix less likely, reducing so the daily release of vancomycin.

The results of our investigations conducted into the femoral condyle of rabbits show that the release of vancomycin occurred *in vivo*. From the residual concentrations into plaster pellets extracted after increasing times of implantation into bone tissues it appears that the elution of the drug decreased with time, and would suggest that a drug release could occur again. In Table 1 we can see that whatever the type of implant coated or not with the polymer, concentrations of vancomycin exceeding MIC were obtained around the implant in all bone tissues. The bone penetration of the antibiotic closely depended on the coating depth. The uncoated implants induced osseous penetration of drug during one week; the two layers of $PLA_{45}GA_{10}$ allowed to extend the duration of the antimicrobial activity for two weeks and the six layers of $PLA_{45}GA_{10}$ allowed to keep therapeutic concentrations for four weeks after a delay of one week. So, the simultaneous implantation of these three types of pellets would be able to produce a burst effect during the first week and a sustained release during at least one month.

The serum concentration of an antibiotic which is important for evaluating systemic side effects, reached a peak concentration (< to toxic levels) three hours after implantation of uncoated implant but remained at a low level thereafter. When coated implants were used, vancomycin was always undetectable in the serum. These data suggest that a high concentration of antibiotic can be obtained at the bone-implant interface for a prolonged period (during one month) with the depot administration of this drug in plaster of Paris implants coated with $PLA_{45}GA_{10}$. The parenteral administration of antibiotics for four to six weeks is usually recommended for the treatment of chronic osteomyelitis. So, the above results obtained by local antibiotherapy suggest that the simultaneous use of uncoated and coated implants of plaster of Paris fulfils this condition. Although it is necessary to perform similar experiments using bones in pathological states, the biocompatibility and biodegradability of the carrier, the low serum concentrations of the drug reducing the risks of systemic side effects and its controlled delivery suggest the potential use of this system for the clinical treatment of bone and joint infections. Its application may be further widened by changing both antibiotic and loading dose as well as the ratio of PLA and PGA to control the drug release and the biodegradability of the carrier and by adding osteoinductive agents (bone morphogenetic protein, demineralized bone matrix, marrow cells or cultured chondrocytes) to promote osteogenesis.

REFERENCES

1. D.S. Kahn and K.P.H. Pritzker, The pathophysiology of bone infection, *Clin. Orthop.* **96:**12-19 (1973).
2. F.A. Waldvogel, G. Medoff and M.N. Swartz, Osteomyelitis: a review of clinical features, therapeutic considerations and unusual aspects, *N. Engl. J. Med.* **282:** 198-322 (1970).
3. T.J. Marrie and J.W. Costerton, Mode of growth of bacterial pathogens in chronic polymicrobial human osteomyelitis, *J. Clin. Microbiol.* **22:**924-933 (1985).
4. K.J. Mayberry-Carson, B. Tober-Meyer, J.K. Smith, D.W Jr Lambe and J.W. Costerton, Bacterial adherence and glycocalyx formation in osteomyelitis experimentally induced with *Staphylococcus aureus, Infect. Immun.* **43:**825-833 (1984).
5. J.C. Nickel, I. Ruseska, J.B. Wright and J.W. Costerton, Tobramycin resistance of *Pseudomonas aeruginosa* cells growing as a biofilm on urinary catheter material, *Antimicrob. Agents Chemother.* **27:**619-624 (1985).
6. B.B. Hall and R.H. Fitzgerald, The pharmacokinetics of penicillin in osteomyelitis canine bone, *J. Bone Joint Surg.* **65A:** 526-532 (1983).
7. Ch. Jr Organ, The utilization of massive doses of antimicrobial agents with isolation perfusion in the treatment of chronic osteomyelitis, *Clin. Orthop.* **76:**185-192 (1971).
8. H.N. Oguachuba, Use of instillation-suction technique in treatment of chronic osteomyelitis, *Acta Orthop. Scand.* **54:**452-458 (1983).
9. C.R. Perry, J.K. Ritterbusch, S.H. Rice, K. Davenport and R.E. Burdge, Antibiotics delivered by an implantable drug pump. A new application for treating osteomyelitis, *Am. J. Med.* **30 Suppl. 6B:**222-227 (1986).
10. K. Klemm, Die behandlung chronischer knocken infektionen mit gentamycin PMMA-kelten and-Kugelen. In: Gentamycin PMMA Kette, Gentamycin PMMA Kugelin Symposium, H. Contzen, Ed., VLE Verlag, Munchen pp. 20-25 (1977).
11. W.R. Murray, Use of antibiotic-containing bone cement, *Clin. Orthop.* **190:**89-95 (1984).
12. T.N. Gerhart, R.D. Roux, G. Horowitz, R.L. Miller, P. Hanff and W.C. Hayes, Antibiotic release from an experimental biodegradable bone cement, *J. Orthop. Res.* **6:**585-592 (1988).
13. Y. Ikada, S.H. Hyon, K. Jamshidi, S. Higashi, T. Yamamuro, Y. Katutani and T. Kitsugi, Release of antibiotic from composites of hydroxyapatite and poly(lactic acid), *J. Controlled Rel.* **2:**179-186 (1985).
14. S.S. Sampath, K. Garvin and D.H. Robinson, Preparation and characterization of biodegradable poly (L-lactic acid) gentamicin delivery systems, *Int. J. Pharm.* **78:** 165-174 (1992).
15. G. Wei, Y. Kotoura, M. Oka, T. Yamamuro, R. Wada, S.H. Hyon and Y. Ikada, A bioabsorbable delivery system for antibiotic treatment of osteomyelitis - The use of lactic acid oligomer as carrier, *J. Bone Joint Surg.* **73B:**246-251 (1991).
16. P. Ylinen, Filling of bone defects with porous hydroxyapatite reinforced with polylactide or polyglycolide fibres, *J. Materials Sci. M. M.* **5:**522-528 (1994).
17. H. Oghushi, V.M. Goldberg and A.I. Caplan, Repair of bone defects with marrow cells and porous ceramic - Experiments in rats, *Acta Orthop. Scand.* **60:**334-339 (1989).
18. Y. Yamazaki, S. Oida, Y. Akimoto and S. Shioda, Response of the mouse femoral muscle to an implant of a composite of bone morphogenetic protein and plaster of Paris, *Clin. Orthop.* **234:**240-249 (1988).
19. J.A. Setterstrom, T.R. Tice and E. Jacob, Chemotherapeutic treatment of bacterial infections with an antibiotic encapsulated within a biodegradable polymer matrix, Patent WO 91/13595 (1991).
20. K. Yamamura, H. Iwata and T. Yotsuyanagi, Synthesis of antibiotic loaded hydroxyapatite beads and *in vitro* drug release testing, *J. Biomed. Mater. Res.* **26:**1053-1064 (1992).
21. X. Zhang, U.P. Wyss, D. Pichora, M.F.A. Goosen, Biodegradable controlled antibiotic release devices for osteomyelitis. Optimization of release properties, *J. Pharm. Pharmacol.* **46:**718-724 (1994).
22. H. Dreesman, Ueber Knochenplombierung, *Beitr. Klin. Chir.* **9:**804-810 (1892).
23. V. Dacquet, A. Varlet, R.N. Tandogan, M.M. Tahon, L. Fournier, F. Jehl, H. Monteil and G. Bascoulergue, Antibiotic-impregnated plaster of Paris beads. Trials with teicoplanin, *Clin. Orthop.* **282:**241-249 (1992).

24. D. Mackey, A. Varlet and D. Debeaumont, Antibiotic loaded plaster of Paris pellets: an *in vitro* study of a possible method of local antibiotic therapy in bone infection, *Clin. Orthop.* **167**:263-268 (1982).
25. B. Mousset, M.-A. Benoit, C. Delloye, R. Bouillet and J. Gillard, Biodegradable implants for potential use in bone infection : *in vitro* study of antibiotic-loaded calcium sulfate, *Int. Orthop.* **19**:157-161 (1995).
26. B. Mousset, M.-A. Benoit, R. Bouillet and J. Gillard, Le plâtre de Paris : un vecteur d'antibiotiques pour le traitement des infections osseuses, *Acta Orthop. Belg.* **59**:239-248 (1993).
27. L.F. Peltier, The use of plaster of Paris to fill defects in bone, *Clin. Orthop.* **21**:1-31 (1961).
28. A. Petrova, Gipsfüllung von knochenhöhlen bei osteomyelitis, *Zentralorg. Ges. Chir.* **43**:885 (1928).
29. R. Bouillet, B. Bouillet, N. Kadima and J. Gillard, Traitement de l'ostéomyélite chronique en milieu africain par implants de plâtre imprégné d'antibiotiques, *Acta Orthop. Belg.* **55**:1-11 (1989).
30. J. Evrard, Expérience clinique des linguettes de plâtre chargées d'antibiotique dans le traitement des ostéites chroniques, *Orthop. Traumatol.* **3**:59-64 (1993).
31. I. Sulo, Granules de plâtre à la gentalline dans le traitement de l'infection osseuse, *Rev. Chir. Orthop.* **79**:299-305 (1993).
32. A. Varlet and P. Dauchy, Billes de plâtre de Paris aux antibiotiques dans le traitement de l'infection osseuse. Nouvelles associations plâtre-antibiotiques, *Rev. Chir. Orthop.* **69**:239-244 (1983).
33. D.K. Kuechle, G.C. Landon, D.M. Musher and P.C. Noble, Elution of vancomycin, daptomycin, and/ amikacin from acrylic bone cement, *Clin. Orthop.* **264**:302-308 (1991).
34. M.-A. Benoit, B. Mousset, R. Bouillet and J. Gillard, Antibiotic-loaded biodegradable pellets for bone infection. Long term treatment, *Proceed. J. Bone Joint Surg.*, in press.
35. D.E. Cutright and E.E. Hunsuck, Tissue reaction to the biodegradable polylactic acid suture, *Oral Surg.* **31**:134-139 (1971).
36. R.K. Kulkarni, K.C. Pani, C. Neuman and F. Leonard, Polylactic acid for surgical implants, *Arch. Surg.* **93**:839-843 (1966).
37. J. Mauduit, N. Bukh and M. Vert, Gentamycin / poly(lactic acid) blends aimed at sustained release local antibiotic therapy administered per-operatively. III - The case of gentamycin sulfate in films prepared from high and low molecular weight poly(DL-lactic acids), *J. Controlled Rel.* **25**:43-49 (1993).
38. M. Vert, P. Christel, P. Chabot and J. Leray, Bioresorbable plastic materials for bone surgery. In : Macromolecular Biomaterials, G.W. Hashings and P. Ducheyne, Eds, CRC Press, Boca Raton, pp. 119-142 (1984).
39. D.C Grove, A.W. Randall, Assay methods of antibiotics, Medical Encyclopedia Inc. New-York (1955).

THE USE OF BIORESORBABLE IMPLANTS IN ORTHOPAEDICS

A. Mazhar Tokgözoğlu

Clinical Associate Professor of Orthopaedics and Traumatology
Department of Orthopaedics and Traumatology
Hacettepe University Faculty of Medicine
Hacettepe 06100
Ankara, Turkey

Internal fixation of fractures has become the treatment of choice in many instances. With internal fixation, fragments of the broken bone are surgically exposed through an incision and adapted to each other manually. Following this, the fracture fragments are attached to each other with metallic implants. The purpose of this approach is to achieve a rigid fixation, making long periods of external fixation obsolete while the patient is encouraged to actively use the injured extremity decreasing morbidity caused by the fracture. This also decreases the cost of a fracture to the society by returning the patient to the activities of daily living as soon as possible. The disadvantage of this approach is that in most instances removal of these implants requires a second operative procedure, anaesthetic application, surgical incision that also increases the cost to the society.

If an implant with sufficient strength to replace metallic fixation devices that does not require surgical removal can be developed many problems of internal fixation may be solved. Bioresorbable materials have been used in surgery since 175 AD for this goal. The first resorbable material was a resorbable gut suture developed by Galen. However, Orthopaedic application of bioresorbable materials is new. PLA (polylactic acid), PGA (polyglycolic acid), and PHBA (poly-beta-hydroxybutyric acid) implants were used to fix tibial osteotomies in rabbits with success[1]. The first extremity fracture in a human was performed in 1985 in Finland where malleolar fractures were fixed using Polygalactin 910®, a copolymer of PGA and PLA. In 1987, Johnson & Johnson introduced the PDS pins into the United States[2]. Initial problems with strength and biocompatibility were addressed by improvements in the materials. Since then bioresorbable implants have been especially used frequently in foot surgery.

In 1994, a randomised prospective study of malleolar fractures compared PLA screws to metallic screws and showed that biodegradable screws were a reasonably safe and effective alternative[3]. New devices are being tried and are gaining approval by FDA and other regulatory institutions.

Clinical studies have demonstrated that these implants are completely resorbed after implantation in bone. The implant is gradually replaced by bone as it is resorbed. The PGA based implants are degraded to glycine or oxalic acid. PLA implants degrades to pyruvic

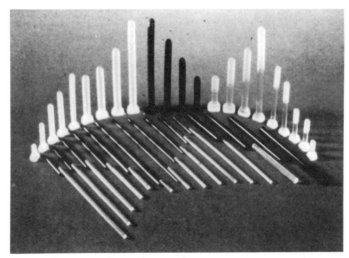

Figure 1. Various types of screws and Kirschner wire type devices that are used is demonstrated. There are many sizes and types of screws available.

acid, which is consumed in the tricarboxalic acid cycle. Eventually the breakdown products of these implants are excreted in urine, faeces, or respiration. Improvement of the manufacturing techniques, sterilisation procedures, reinforcement of the materials has made the use of these materials safer[4]. The problems of sterile abscess formation, celulitis at the application site in early clinical trials have been decreased with material improvements. Currently there are many different implants available to the use of the Orthopaedic community. The most commonly used implants are Kirschner wire type devices, bone screws, suture anchors, devices to attach soft tissues to the bone (Figures 1,2).

ADVANTAGES OF USING RESORBABLE DEVICES IN ORTHOPAEDIC SURGERY

Elimination of the Second Surgical Procedure for Removal

The main reason for using resorbable materials is to avoid taking a healthy patient back to surgery for removal of a fixation device. In long term, by eliminating the cost of a second surgical procedure the savings to general public is large. In addition, the patient is not put under the agony and stress of a surgical procedure. This also eliminates potential problems due to implant removal such as infections, anaesthesia complications, and absence from work due to recovery.

Decreased Load Sharing

Prolonged presence of implanted plates may cause bone resorption and weakening under their surface due to their load sharing effect. This may cause weakness to the bone and when the implant is eventually removed, predispose the patient to a second fracture and bone pain. Bioresorbable plates with gradual resorption capabilities may lessen this problem.

Multi-Tasking (Drug Delivery)

During the manufacturing process certain materials, such as antibiotics or bone morphogenetic proteins may be added to the bioresorbable internal fixation devices. This

198

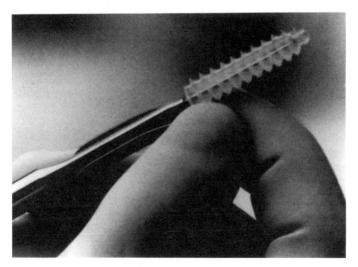

Figure 2. A new PLA device for fixation of the tendon graft for anterior cruciate ligament of the knee is demonstrated. The device is used with the specially designed screw driver for application using an arthroscope.

property potentially can enhance to an extent healing of certain fractures. Such as in open comminuted fractures bone healing can be enhanced while local infection is cured. Another application is filling bone defects in chronic osteomyelitis while the bone is sterilized by antibiotic delivering materials.

Cost-effectiveness

By eliminating the necessity of hardware removal, bioresorbable implants can decrease cost of treatment and avoid complications. By eliminating the implant removal operation significant cost reduction is possible with eliminating the costs of following procedures: second admission, second anaesthetic, surgeon fees, operating room costs, hospital stay and absence from work due to the second surgical procedure, treatment of potential complications.

Absence of metal toxicity

While risks of long term implantation is not well known, the potential of carcinogenesis and toxicity due to corrosion of metallic implants is present when using metal based implants. Wear creates micro particles of metal implants, which are phagocytosed by the macrophages. Bioresorbable implants negate this risk since they do not contain nor create metallic particles.

DISADVANTAGES OF USING BIORESORBABLE DEVICES IN ORTHOPAEDIC SURGERY

Cost

One of the current drawbacks of bioresorbable implants is the cost of the implant. With the current cost of manufacturing, bioresorbable implants are about 15 times more expensive than a similar metallic implant. Mass production and advances in manufacturing techniques will increase cost efficiency in the near future.

Strength

The strength and stiffness of bioresorbable implant is less than those of metallic implants. The most common and effective force in fixation of bone fragments is the compression force created between the two segments. To achieve rigid fixation, which is required for bone fixation, significant compression between both fragments is needed. When used as a screw bioresorbable implants provide less compression compared to metallic screws. The treads and shaft of PLA screws are still not as strong as metallic screws. To achieve good fixation rigid adherence to the technique is essential. The bone should be preliminary fixed and drilling followed by tapping is necessary before screw insertion. This property of bioresorbable implants currently limits their use to low load and low yield situations such as cancellous and small bone fixation in adults, and fracture fixation in children.

Foreign Body and Delayed Tissue Reactions

Some patients (8%) have experienced foreign body reactions at the site of application requiring aspiration. Symptomatic swelling has been reported as a late reaction in some cases. This has been treated with debridement. While these reactions have not adversely effected bony healing, their presence contradicts the purpose of using such materials. Sinus formation due to this debris has been reported[5]. Early removal of the device has been reported in some instances. These problems were reported mostly in early phases of application. Manufacturing processes such as orientrusion has decreased the incidence of this problem. Perhaps this problem is related to fast decay of the implant, which overcome the ability of local mechanisms to absorb the byproducts.

Radiolucency

For medical or legal reasons many users would like to see implanted materials on radiographs. This is necessary for demonstrating the proper use of the device in patients. Current implants do not contain radio-opaque material, which will allow radiographic demonstration of the implant. For users or indications that require a radiographic demonstration of the implant, the fact that bioresorbable implants are radiolucent may be a disadvantage.

Shelf Life

Being a biological plastic polymer, bioresorbable implants have a shelf life. This fact increases their cost and decreases their applicability. Efficient stock control is difficult and unpredictable. With improvements in technology, newer implants have a longer shelf life.

Deterioration with Manipulation

Being high molecular weight polymers bioresorbable implants can not withstand any physical or chemical manipulation especially resterilization. Water and moisture have deleterious effect on these materials. If repositioning or remanipulation is necessary during use, in most cases another implant may have to be used, which will increase the overall cost of the operation.

APPLICATION OF BIORESORBABLE IMPLANTS IN ORTHOPAEDIC SURGERY

Adults

Bioresorbable implants have been successfully used in metaphyseal fractures of ankle and small bones. However, in about 10% of cases spontaneous drainage or sterile abscess formation has been reported. In a study using *"orientruded"* PLA screws no problems were reported. The orientrusion process enhances the strength of the material and allows a longer clearance of PLA probably improving the tissue tolerance and the body's handling[4]. Important applications of bioresorbable implants in adults are osteotomies of the small bones of the foot. The most common use of bioresorbable implants in adults is metatarsal osteotomies in hallux valgus surgery. Since these bones have fast healing rate and do not require strong forces for fixation, bioresorbable implants provide an important fixation option.

Children

In many children's fractures Kirschner wires are used since they are relatively easy to remove and provide fixation that is enough in children. However, their removal may still cause important problems. Kirscher wires are also notorious for their migration. Bioresorbable pins may replace these implants since they do not require removal. Since children's fractures heal much faster than adults, small size PGA rods have been used with success in supracondylar fractures, shear fractures of the medial epicondyles, and small bones of the hand and foot. Larger PLA rods and screws have been used in some cases of osteotomies of the cancellous pelvic bone[5,6].

PLA rods have been used successfully in fractures involving or crossing the physeal growth plates. Animal and human studies have demonstrated that growth disturbance is not a problem. With their biological deterioration, PLA screws provide rigid fixation in the early phases of healing and later loose their fixation strength avoiding compression across the growth plate (Figure 3).

By avoiding the implant using bioresorbable devices, we can save children from the unhappy experience of a second surgical procedure. These devices are still not strong enough to use in fixation of large bones in children.

OTHER POSSIBLE ORTHOPAEDIC USES OF BIORESORBABLE IMPLANTS

Sports Medicine

Bioresorbable devices have been used in fixation of the tendon graft used in reconstruction of the anterior cruciate ligament (ACL) of the knee. Large size screws or tacks have been tried for this purpose. Early implants did not provide the fixation strength that is provided by metallic devices and caused some synovitis. New generation devices provide fixation similar to metallic devices and MRI studies of these improved implants has not shown adverse reactions. Bioresorbable implants appear to be an important addition to ACL surgery[5] since most current fixation methods require the use of metallic fixation devices that prevent MRI imaging of the reconstructed graft (Figure 4).

Figure 3. Photograph of a specimen from an experimental study. Void spaces left over from two absorbed screws is seen. The fixation screws have not caused any damage to the distal physeal line of the femora of a rabbit.

FUTURE ORTHOPAEDIC USES OF BIORESORBABLE IMPLANTS

Absorbable clips for meniscus repair in knee surgery is probably one of the most promising uses of bioresorbable implants. They will be easy to use attachment devices that can be used via an arthroscope. Drug delivery systems for osteomyelitis is another promising application for the near future. Creating Orthopaedic fixation devices with antibiotic or anti-infective properties is an exciting concept that is being sought. Defect management and sterilization of cavitary defects in chronic osteomyelitis is an important problem in treating chronic bone infections. By filling cavitary bone defects with bioresorbable implants that release antibotics, one can possibly treat this difficult problem. Other innovative potential uses are special strong screws for acetabular cup fixation in total hip arthroplasty. The screws in current use demonstrate the potential problem of backing up on the polyethylene insert and cause wear. By using resorbable screws this potential wear problem may be eliminated. Unfortunately, current screws are not strong enough to withstand the torque forces required for this type of fixation.

Bone graft substitutes to reconstruct large defects is another exiting potential use for bioresorbable implants. Large bone defects that are caused by trauma is a challange to orthopaedic surgeons often requiring multiple and difficult operations. Bioresorbale implants may be used[5] as scaffolds for bone grafts, defect fillers loaded with bone morphogenetic proteins to stimulate bone reconstitution. Anti-adhesion membranes are under investigation to prevent adhesion, which is an important problem in tendon repairs. Ligament augmentation devices[5] have been tried without much success in anterior cruciate ligament of the knee surgery. Newer devices are being tried in lower demand situations, such as ankle ligaments.

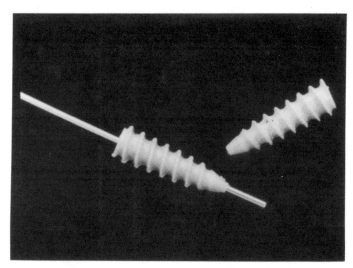

Figure 4. A new implant under development for fixation of the tendon graft used in reconstruction of the anterior cruciate ligament of the knee. Although this device is radiolucent and not visible on a conventional radiograph, it does not interfere with magnetic resonance imaging, This allows good visualization of the graft for evaluation of the results of the surgery. Otherwise an arthroscopic intervention which is a surgical procedure would be necessary to evaluate the graft.

Bioresorbable implants definitely have a wide area of application in Orthopaedic Surgery. Future research will bring even more innovative devices to the use of Orthopaedic surgeons.

REFERENCES

1. S. Vainionpaa, K. Vihtonen, M. Mero, H. Patiala, P. Rokkanen, J. Kilpikari, P. Tormala, 1986, Biodegradable fixation of rabbit osteotomies, *Acta Orthop. Scand.* **57**:237-239.
2. D.C: Tunc, M.W. Rohowsky, B. Jadhav, W.B. Lehman, A. Strongwater, F. Kummer, 1987, Body absorbable osteosynthesis devices, in: *Advances in Biomedical Polymers*, (C.G. Gebelein, ed.), pp. 87-99, Plenum Press, New York.
3. R.W. Bucholz, S. Henry, M.B. Henley, 1994, Fixation with bioabsorbable screws for the treatment of fractures of the ankle, *J Bone Joint Surg* **76A**:319-324.
4. D.C. Tunc, B. Jadhav, 1990, Development of absorbable, ultra high strength poly(lactides), in: *Progress in Biomedical Polymers*, (C.G. Gebelein, R.L. Dunn, eds.), pp. 239-248, Plenum Press, New York.
5. R.D. Blaiser, R. Bucholz, W. Cole, L.L. Johnson, E.A. Mäkelä, 1996, Bioresorbable implants: applications in Orthopaedic surgery, *Instr Course Lec* **46**:531-546.
6. R.K. Fraser, W.G. Cole, 1992, Osteolysis following biodegradable pin fixation of fractures in children, *J Bone Joint Surg* **74B**:929-930.

THE DEVELOPMENT OF A NOVEL COMPOSITE BONE SUBSTITUTE FOR USE IN DRUG RELEASE

Murat Burak Yaylaoğlu[1], Petek Korkusuz[3], Ülken Örs[3], Feza Korkusuz[2] and Vasıf Hasırcı[1]

[1]Biotechnology Research Unit
Department of Biological Sciences
Middle East Technical University
Ankara, Turkey
[2]Medical Center
Middle East Technical University
Ankara, Turkey
[3]Department of Histology and Embryology
Faculty of Medicine
HacettepeUniversity
Ankara, Turkey

INTRODUCTION

A bone defect might regenerate more predictably if a stromal substitute were implanted to provide a framework for organization of the osteons. By providing a bone defect with a stromal substitute containing spaces morphologically compatible with osteons and their vascular interconnections, a partnership between biomaterials and skeletal regeneration may be encouraged.

Introduction of an antibiotic or another bioactive agent could help recovery or the repair process through local delivery of these agents. The objective in this study was to constitute a composite material consisting of an inorganic phase of calcium phosphate which is osteoinductive, and an organic phase similar to that of bone, gelatin. Calcium phosphate was introduced into the gelatin membrane (after formation in situ outside the membrane) or by formation within the membrane via alternating introduction of the calcium and phosphate ions to constitute the mineral phase. An antibiotic, gentamicin sulfate, was also introduced into the composite to act as a remedy against bone infection. Following preparation of the implant, kinetics of antibiotic release and the organization and composition of the implant was studied. The composite was then implanted into rabbit tibia, and antibiotic release and implant biocompatibility were studied.

MATERIALS AND METHODS

Preparation of Calcium Phosphate In Situ

Equal volumes of calcium and phosphate containing Tris buffers were mixed at room temperature. The liquid immediately became turbid and after standing for 15 minutes, was centrifuged (5000 rpm for 10 min. at 4°C), vacuum dried and ground in a ceramic mortar before use.

Gelatin Membrane Preparation

Gelatin (Oxoid) was weighed, added into water and then heated to 60°C to prepare the gelatin solution (15, 20, 25 and 35% (w/v)). In some tests calcium or phosphate containing Tris buffers were used instead of distilled water. The gelatin solution was then poured in a Petri dish and cooled down to room temperature

Composite Preparation by Calcium Phosphate Growth in Gelatin Membrane

Gelatin membrane (15, 20, 25 and 35 % gelatin (w/v)) was immersed overnight in calcium containing Tris buffer washed with distilled water and then immersed overnight in phosphate containing Tris buffer. A milky layer immediately started to form on the surface of the membrane, intensity of which increased with time. The gel was then washed in distilled water, crosslinked by immersion in glutaraldehyde (8 %) and then dried in a vacuum oven overnight. The composition and morphology of the resultant membrane was examined with infrared spectroscopy (IR), scanning electron microscopy (SEM) and x-ray diffraction spectroscopy (XDS).

Gentamicin Loaded Composite Preparation By Using Calcium Phosphate Containing Gelatin Membranes

Gelatin required to prepare 15 % (w/v) solution, gentamicin sulfate 75 mg/mL in powder form and calcium phosphate required to yield 75 mg/mL were weighed and added to distilled water. The mixture was warmed in a water bath at 60°C for 1 h and vortexed before pouring into a Petri plate. After cooling and solidification, rectangular implants (2 x 10 x 10 mm) were cut out and air-dried overnight.

In Vivo Experimentation

Implantation of Gelatin Membranes into the rabbits; eight three-month-old female local albino rabbits (New Zealand Wild Type) were used. After anaesthesia a window of 2 x 10 x 10 mm size was prepared at the metaphyseal-diaphyseal junction using an electric powered torque controlled drill. The cortex was drilled at each side of the rectangular area to prevent further cracks. The implants were sized according to the dimension of the cortical window and implanted into the bone. A pair of rabbits were sacrificed on the first, second, third and fourth weeks of implantation. Both legs of one of the rabbits were transferred into 10 % formaldehyde containing tubes and used in histological examination while the tibia of the second animal was used to determine the *in vivo* drug release.

Microbiological Assay of Antibiotic

As gentamicin sulfate does not absorb in the UV or visible regions, a microbiological assay using *(E. coli)* was required for its quantification[1].

In Situ Drug Release

A calcium phosphate and gentamicin sulphate loaded (75 mg/mL) gelatin membrane (2 x 10 x 10 mm) was placed in phosphate buffer solution (PBS) in a 10 mL flask and stored at room temperature. PBS was changed initially daily, then weekly and bi-weekly.

In Vivo Drug Release

Gentamicin released from the implants *in vivo* was calculated from the mass balance, by determining the amount of gentamicin sulphate still remaining within the implant after it was removed from the rabbit tibia. The amount of gentamicin that leached out of the implant was determined through a microbiological assay.

Histology

All specimens were decalcified in a solution of 10 percent formic acid for 1 to 2 weeks. The decalcified specimens were immersed in sodium sulphate overnight and immersed again in the appropriate fixative. All specimens were dehydrated in a graded series of ethanol and were then embedded in paraffin. Five micrometer thick longitudinal sections of the decalcified bone specimens and the implant were prepared, and stained with Hematoxylin-eosin, Mallory trichrome and Masson trichrome. Photomicrographs were obtained by an Olympus BH-7 light microscope.

RESULTS AND DISCUSSION

Infrared Spectroscopy (IR)

The peak assignments for the spectra of hydroxyapatite is reported in literature[2] as: OH absorption: 630 cm^{-1}; PO_4^{-3} absorption: 565, 601, 962, 1032, 1100 cm^{-1}; adsorbed water: 1648, 3000, 3600 cm^{-1}.

In Figure 1, the IR of the gelatin membrane, calcium phosphate, calcium phosphate loaded gelatin membrane and membrane containing crystalized calcium phosphate are presented. The distinct contribution of gelatin is the absorbance due to the amide linkages. Amide I, II and III appear at 1640, 1550 and 1350 cm^{-1}, all of which are observed in the spectra of gelatin and the composite, but not in the spectra of calcium phosphate. The major difference is the appearance of a phosphate peak at ca. 1120 cm^{-1} in the gelatin membrane within which calcium phosphate was gradually formed. Since this peak is not observed when calcium phosphate is loaded as granules during the setting of the other type of membrane this peak can be taken as an indicator of a close interaction between the organic component and the newly formed crystal. Thus, a more intimate contact (mimicking bone) has been established.

Scanning Electron Microscopy

When the gelatin membrane is used as a crystallisation scaffold in the formation of calcium phosphate crystals, the rosette structures shown in Figure 2 are obtained (the surface). Especially distinct is the presence of a skin layer and when this is examined at a higher magnification, the same rosette shaped formation of calcium phosphate crystals is observed here too. These structures are absent in the calcium phosphates prepared by

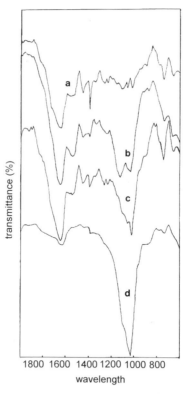

Figure 1. IR spectra of free medium a: Gelatin b: Calcium phosphate crystalized and c: loaded into gelatin. d: Calcium phosphate.

Figure 2. SEM of calcium phosphate crystalized within gelatin membrane (left). SEM of the cross-section of the gelatin membrane (right).

crystallisation in the absence of a support or a scaffold revealing the influence of the gelatin matrix as a template. This is another confirmation of the close association and interaction revealed by the IR data.

XDS Study

XDS can determine the amount of an individual element with an atomic number greater than six. This study is conducted to determine the ratios of calcium to phosphate and to compare the calcium phosphate composition to that of hydroxyapatite. It was observed in the XDS study that the commercial hydroxyapatite had a calcium to phosphorus molar ratio of 1.44 and the calcium phosphate crystal formed in gelatin membrane had a ratio of 2.31 (Table 1). The in situ calcium phosphate has a similar value (2.33). Since the theoretical value for hydroxyapatite is 1.67, the crystals formed in this study are then "calcium rich". Thus, the calcium phosphate crystals formed in this study will be more readily soluble in the biological medium[3] because of this and it might be advantageous because as they dissolve bone will grow in and a more intimate tissue-material contact will be established.

Table 1. Quantitative XDS analysis for calcium to phosphorus ratio

| | | (Ca:P ratio) | |
		Molar	Weight
Sample	Commercial Hydroxyapatite	1.44	1.86
	Calcium Phosphate Crystal Growth On Surface of Gelatin Gel	2.31	2.98
	Calcium Phosphate Prepared in Free Medium	2.33	3.02
Theoretical	Hydroxyapatite $Ca_{10}(PO_4)_6(OH)_2$	1.67	2.16
	Tetracalcium Phosphate $Ca_4P_2O_9$	2.00	2.49
	Tricalcium Phosphate $Ca_3(PO_4)_2$	1.50	1.94

One disconcerting observation is that the value of 1.44 was obtained with the commercial HAp. It is either because the crystals are not hydroxyapatite as claimed or the XDS is off by 0.23 units or ca 15 %. If the latter is correct, the values obtained, upon correction, reach 2.50, which are more distinctly higher than expected from HAp.

In Situ Drug Release

The amount of gentamicin sulphate released from the implants *in situ* was determined through a microbiological-assay using *E.coli* K12 [1].

Initially entrapment efficiency of the membranes was determined. It was found that the 15 % membrane retained 2.46 % of the implant antibiotic where as the 25% membrane retained a much higher value (27.61%). The major difference between these two is the concentration of the gel within a given volume (or gel density) and this increased presence of the gelatin fibers prevents, due to diffusional restrictions and due to increased interaction, the leaching of the antibiotic during the various preparation steps. This was also expected to influence the release roles for the same reason.

In situ release of gentamicin reveals a burst in the initial stages (Figure 3). Release from 15 % gelatin membrane is achieved at a much lower level because of its 10 fold lower antibiotic content. In both cases, a release duration of more than 60 days is obtained. In order to reveal the kinetics, logarithm of the released drug are plotted against time (Figure 3). Here it is seen that if the initial burst is excluded, the release behavior can be explained by a first order kinetics. The other observation is that the release from the 15 % gelatin membrane is much more rapid than it is from the 25 % gelatin membrane. This was expected from its lower gelatin and therefore higher water content and density.

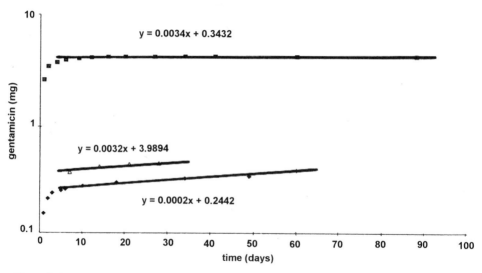

Figure 3. *In situ* (above: 25% gelatin, bottom: 15% gelatin) and *in vivo* (middle: 15% gelatin) gentamicin release, Log gentamicin vs. time.

In Vivo Drug Release

In Figure 3, *in vivo* gentamicin release behavior of the implants are also presented. The values are calculated from the antibiotic remaining in the implants and from the mass balance.

As it was observed in the *in situ* release, there was an initial burst of antibiotic from the implant under *in vivo* conditions, too. Although the extent of release in the first days was lower than the *in situ* application, complete release was obtained in a much shorter duration (30 days in *in vivo* against 80 days *in situ*). The release kinetics appears to be first order as it was in the *in situ* case (Figure 3). An initial burst could actually be quite useful in eradicating the microorganisms at the infection site while the later, lower rate of release, could be effective in maintaining the medium sterile.

The shorter duration in the *in vivo* application could be due to the hydrolytic and enzymatic degradation of the gelatin fibers and resorption of the implant at the implant site.

Histology

One week after implantation into the soft tissue (Figure 4) or into the bone (Figure 5 and 6), whether the implant is loaded with gentamicin or not, did not adversely effect new bone formation. Degradation of gelatin and the penetration of its particles into the soft tissue was an obvious finding. When implanted into the bone, a loose connective tissue surrounded the implant during the first two weeks which then became denser during the third week (Figure 7). Dense connective tissue was replaced by new bone trabecules and cartilage tissue at the same time. The implant did not effect the natural course of periosteal and enchondral new bone formation (Figure 7 and 8). Macrophage reaction towards the implant occurred during the third week (Figure 8) and was more obvious during the fourth week. The delayed onset of macrophages may closely be related to the degradation process of the implant. Enchondral bone formation was the common finding at the end of four weeks (Figure 9). Immature bone converted into mature bone in the implantation site. Numerous macrophages surrounding the implant at four weeks reminding a delayed

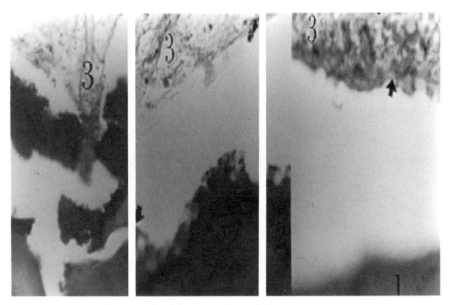

Figure 4. Soft tissue findings of calcium-phosphate-gelatin implantation after mild decalcification. Micrograph presents calcium-phosphate crystals (1) and gelatin (2) in close relation with the connective tissue (3). Mallory trichrome x 10 (left). Gelatin particles (arrow) are detaching from the implant (2) and some particles are present in the connective tissue (3). Mallory trichrome x 40 (center). Gelatin particles (arrows) are obviously located in the connective tissue (3). Mallory trichrome x 40 (right).

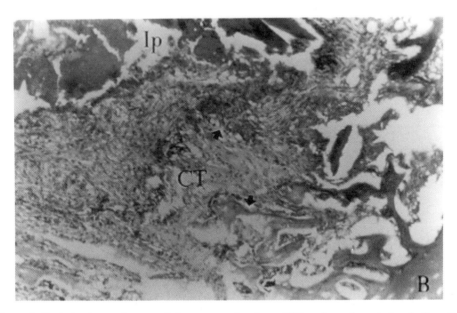

Figure 5. Control micrographs present dense connective tissue (CT) and new bone trabecule formation (arrows) between the implant (Ip) and the mature bone (B) in one week. Masson trichrome x 10

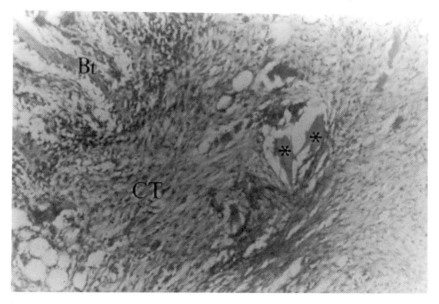

Figure 6. Developing bone trabecules (Bt) are observed with bone residues (asterix) neighbouring the thickened and dense connective tissue (CT). HE x 10

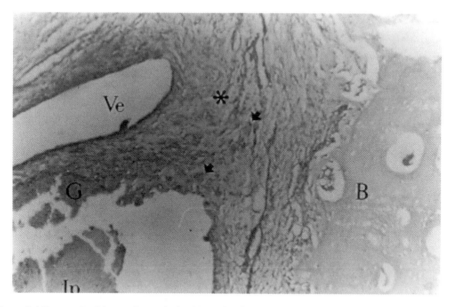

Figure 7. Micrograph of the specimen obtained 3 weeks after implantation. Gelatine particles (G and arrows), belonging to the implant (Ip) are surrounded by the connective tissue. The ventricle (Vc) of the bone (B) is a part of the periosteum (asterix) that is covered over the implant during surgery. HE x 10

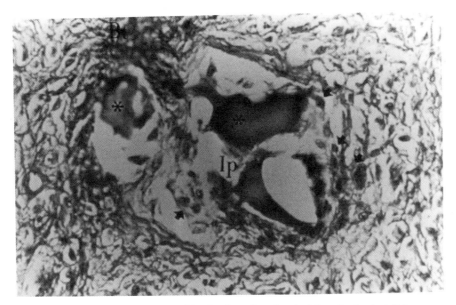

Figure 8. Calcium-phosphate particles (asterix) of various sizes detached from the implant (Ip) are covered by the connective tissue. Thus, new bone formation (Bt) is observed near the implant. The presence of macrophages (arrows) without any major immunological reaction are most probably due to the resorption process. HE x 10

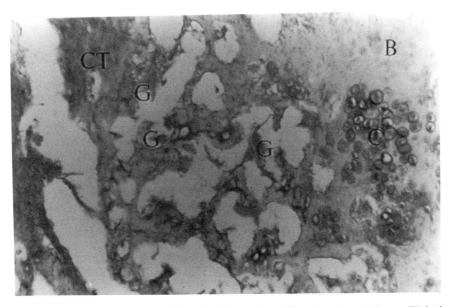

Figure 9. Connective tissue (CT) is partially replaced by cartilage (C) and new bone trabecues (B) in three weeks. Gelatin particles (G) of the gentamicin loaded implant are obviously present in between the bone formation process without causing any adverse effect. HE x 10

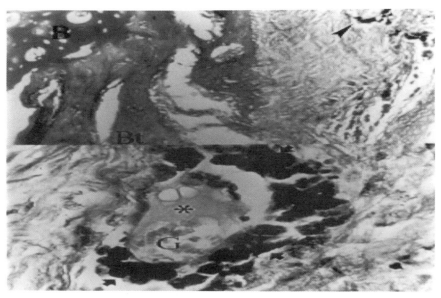

Figure 10. Micrographs of the implantation site in 4 weeks. Immature bone trabecules (Bt) are converting into mature bone (B) tissue in the implantation site. The residues of the implant (arrow) are present in the dense connective tissue. Masson trichrome x 10 (above). Higher magnification of the implant residue of presents calcium-phosphate (asterix) and gelatine (G) particles belonging to the implant surrounded by numerous macrophages (arrows). Masson trichrome x 40 (below).

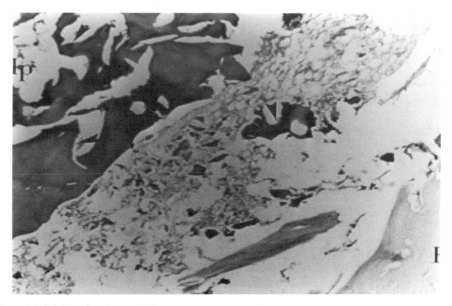

Figure 11. Calcium phosphate particles (arrow) are found in between the implant (Ip) and the mature bone (B). Masson trichrome x 4

214

immunolgical reaction most probably due to the resorption process (Figure 10). Calcium phosphate particles in between the implant and the bone revealed excellent biocompatibility (Figure 11).

CONCLUSION

In this study the implant that was developed to mimic the structure of bone was composed of an organic phase consisting of gelatin and a mineral phase of an apatite-like calcium phosphate. The calcium phosphate crystals had a much higher calcium to phosphorous ratio than HAp implying that it is more erodible[3].

Drug release from the implant was sufficiently long (more than 6 weeks) proving the potential for use in the local administration of an antibiotic for treating diseases like osteomyelitis where blood flow is uncertain and systemic application of the antibiotic requires significantly higher doses.

Histological findings reveal that both, gentamicin loaded and non-loaded calcium phosphate-gelatin implants are highly biocompatible. Both implants allow new bone formation as presented by the Masson trichrome staining method. Gelatin and calcium phosphate particles penetrate into the connective tissue where they are degraded into smaller particles and are surrounded by the macrophages of the tissue. A mild foreign body reaction is observed in both the gentamicin sulphate containing and free groups throughout the experiment. This foreign body reaction seems to be launched against the gelatin component of the implant. The calcium phosphate particles that are separated from the gelatin, on the other hand, are not surrounded by macrophages. Though the immunological reaction is still present after 4 weeks, maturation of the newly formed bone is not inhibited.

In conclusion it can be stated that the calcium phosphate-gelatin implant loaded with gentamicin has a great potential for use in the treatment of various bone diseases like osteomyelitis. This may be an alternative drug release system to the previously defined HAp implant[4], however, comparative studies are essential. Growth of calcium phosphate crystals on a gelatin template also led to detectable differences in the conformation of the crystals encouraging us to do further studies to release its full bone mimicking capability.

REFERENCES

1. D. M. Isaacson and J. Kirschbaum, 1986, Assays of Antimicrobial Substances, (A. L. Demain, and N. A. Solomon, eds.), pp. 410-435, American Society for Microbiology, Washington D.C.
2. M. Shirkhanzadet, M. Azdegan, V. Stack, and S. Schreyer, 1994, Fabrication of pure hydroxyapatite and fluoridated hydroxyapatite coatings by electrolisation. *Materials Letters*, **18**: 211-214.
3. K. Köster, et. al., 1976, Experimental bone replacement with resorbable calcium phosphate ceramic, *Langenbecks Arch. Chir.* **341**: 77-86.
4. Y. Shinto, A. Uchida, F. Korkusuz, N. Araki, and K. Ono, 1992, Calcium hydroxyapatite ceramic used as a delivery system for antibiotics, *J. Bone and Joint Surgery* (Br), **74-B**: 600-604.

USE OF POLYMERS IN UROLOGY

Ali Ergen

Hacettepe University
School of Medicine
Department of Urology
06100 Sıhhiye, Ankara/TURKEY

INTRODUCTION

Non-autologous materials are widely used in the medical field today , and also have a rich and illustrious past. Biomaterials used in urology can be classified as stents, injectables, sphincters and prosthesis. Penile prosthesis and artificial urinary sphincters are totally implantable devices and urinary catheters those with an exit site are transcutaneous devices.

Prosthetic materials must be biocompatible, non-toxic and permanent. Biocompatibility is an utopian state where a biomaterial presents an interface with a physiologic environment without the material adversely affecting that environment or the enviroment adversely affecting the material. The urinary tract is a special environment because the biomaterial is in contact with urine and the urothelium. A biomaterial is defined as a substance with interfaces with tissue at least at some stage of treatment. Biomaterials can be of natural or synthetic origin.

Urinary Stents

Urinary-tract endoprosthesis can be used in the upper or lower urinary tract. Various stents are used in the urethra mostly to treat upper tract obstruction or ureteric fistulas, or to bridge partial defects in the ureteric wall. Alloplastic tubes to replace a gap in the upper urinary tract have been tried with different materials, but these attempts have usually failed due to several reasons.

Urethral stents are used mainly to treat obstruction of the lower urinary tract. They can be classified as permanent stents such as Urolume and ASI stents, and temporary stents such as urethral coils, intraprostatic spirals and intraurethral catheters.

Endoprostheses must have some special futures. Low coefficient of friction is essential for easy insertion. Radioopacity is essential for a correct placement of the stent. Memory is essential to prevent migration. They should have a tensile strength, in order to prevent their breakage and lastly the internal/external diameter ratio should be as low as possible in order to allow for optimal flow conditions inside and around the stent. There is no doubt about the biomaterials' chemical inertness.

Injectables are used mainly for two purposes in urology, endoscopic correction of vesico-ureteral reflux and treatment of intrinsic urethral dysfunction. The ideal bioinjectable material should be nonreactive, non-antigenic, non-migratory, non-erosive and has to be easily placed. Among non-invasive procedures, injected material (polytef, collagen, autologous fat, silicone) can be offered as a relatively simple day case procedure and has potential not only as an effective treatment option but also in reducing the cost for treating incontinence. The best candidates for treatment with injectable materials are those with intrinsic sphincter deficiency with good anatomical support and urodynamic or radiological evidence of failure of the proximal part of the urethra to close at rest in the absence of detrusor contraction. It offers an attractive alternative to surgery especially in the elderly and those unfit for major surgery. Despite encouraging success rates with the endoscopic treatment of urinary stress incontinence and vesicoureteral reflux most of the concerns have focused on both the biocompatibility of the foreign materials and the potential for migration. In particular the migratory potential of polytetrafluoroethylene has made pediatric urologists cautious.

The use of penile prostheses is more detectable. It is intended to provide penile rigidity upon demand in subjects who cannot initiate or maintain an erection sufficient for penetration. Moreover, the ideal implant must increase penile length and girth during intercourse, and allow the penis to return flaccid and non cumbersome otherwise. Many devices have been designed aiming to achieve this goal: rigid, semi-rigid, malleable and inflatable. But the more sophisticated they are, the more fragile and expensive they are too. Data about long-term safety and effectiveness are not available. In spite of the considerable number of implanted devices throughout the world, there are very few studies on patient and partner satisfaction. Most of the complications (infection, erosion, migration, extrusion, mechanical failures) have not been reported. The NIH Consensus Conference December 1992 recommended staging the treatment of erectile dysfunction, beginning with the less invasive ones such as pharmacotherapy, intracavernosal injection and vacuum device.

Total alloplastic replacement of the lower urinary tract will perhaps be an alternative to intestine bladder replacement. The ideal prosthetic bladder should be able to store a large amount of urine at a low pressure and to empty completely without reflux into the kidney. It would have volutional control. It would be made of a totally inert and biocompatible material. It would be surgically easy to implant and be accessible for repairs and replacement if needed. such a prosthesis has never been implanted in a human being. The main cause of failure, in animal experiments, has been the development of fibrous capsule around the prosthesis restricting filling and emptying, leakage of urine at the urethral anastomosis, ureteric obstruction and encrustation leading to hydronephrosis. However solutions are in progress for overcoming these ultimate difficulties.

THE EVOLUTION OF MEMBRANES FOR HEMODIALYSIS

Markus Storr, Reinhold Deppisch, Reinhold Buck, and Hermann Goehl

Renal Care R & D
Gambro Group
D-72379 Hechingen, Germany

INTRODUCTION

Since Willem Kolff, Nils Alwall and other pioneers performed the first regular dialysis treatments on uremic patients in the 1940s, dialysis membranes have been under continual development with respect to improved functionality and biocompatibility. Bulky 'plate kidneys' which were used in the early days of clinical hemodialysis have nowadays been replaced by highly optimized disposable membrane devices containing state of the art synthetic hollow fiber membranes in combination with sophisticated volume-balancing dialysis machines.

Membrane-based treatments for replacement of lost kidney function have gained more and more importance over the past two decades. Acute and chronic renal disease patients have increased from a few thousand in 1970 to 800 thousand in 1996 (figure 1). Out of the 800 thousand 680 thousand are treated with blood purification therapies based upon artificial membranes and about 110 thousand patients receive peritoneal dialysis. Currently 200 thousand people are living with a donor kidney. These numbers explain the great demand for highly efficient and low cost membranes.

The three main treatment modes for membrane-based extracorporeal blood purification are hemodialysis, hemofiltration and hemodiafiltration. In hemodialysis uremic toxins are removed by diffusion across the membrane, whereas in hemofiltration toxins are eliminated by convection, i.e. toxins are removed from the blood by ultrafiltration, the occurring fluid loss is made up by suitable substitution fluids. Hemodiafiltration is a combination of both options, whereby diffusive removal is associated with controlled rates of ultrafiltration. A modern dialyzer has a membrane surface area of about 1.5 square meters. In chronic dialysis with an average treatment time of three to five hours and about 150 treatments per year huge areas of membrane come into contact with the blood of one single patient.

Apart from blood purification in chronic and acute renal failure hemoconcentration after open heart surgery represents a further field of application of such membranes. Per year there is a worldwide demand for 75 million dialyzers for the use in chronic dialysis, 1 million hemofilters for the use in acute dialysis and about 300 thousand hemofilters for the use in heart surgery.

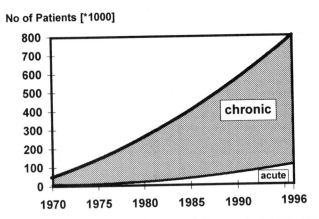

Figure 1. Prevalence of chronic and acute renal disease patients 1970 - 1996.

Considering the complexity and selectivity of the human kidney, the recurrent exposure of blood to foreign surfaces and the recent years with health care budgets diminishing worldwide, the following three basic demands on membrane devices have evolved:

1. An effective and application specific transport behavior is required. Standard hemodialysis (HD) needs high diffusive transport rates for uremic toxins with a molecular weight up to approximately 1,000 D. Hemodiafiltration (HDF) demands high diffusive and convective transport of small and additionally of middle molecules like β2-microglobulin (β2m) and advanced glycosylation endproducts (AGE). In hemofiltration (HF) convection predominates. Both HDF and HF are characterized by a more substantial elimination of substances with a higher molecular weight of up to 40,000 D. However proteins like albumin and other vital blood components should not be removed. Figure 2 gives an overview of the three treatment modes with their desired permeability characteristics and the occurring solute removal mechanisms diffusion and convection.

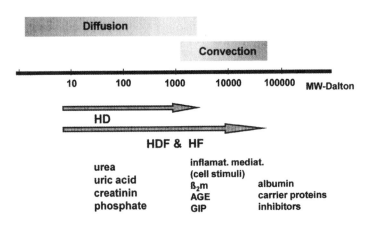

Figure 2. Membrane-based blood purification therapies and removal characteristics.

2. Considering the huge areas of 'non-self' surfaces which come into contact with blood, the demand for high blood compatibility is legitimate. At the blood-material interface the activation of physiological pathways, e.g. the complement system, the coagulation cascade and the contact phase (kallikrein-kinin), have to be minimized. Figure 3 shows the most

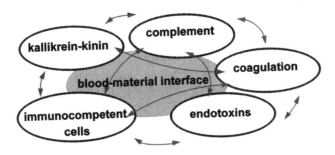

Figure 3. Important activation pathways of blood in contact with foreign surfaces.

important biological mechanisms encountered at the blood-material interface of hemodialysis membranes.

3. Finally, economic aspects have become more and more important in the recent years of limited health care budgets. While raw material and personnel costs have shown a steep increase during the last two decades the selling prices of dialyzers declined by more than seventy percent within the same period. Therefore economic large scale production of high quality membranes is required.

DIALYSIS MEMBRANE MATERIALS

The well known membrane tree first introduced by Michael Lysaght lists most of the currently available membrane types (figure 4). In the search of the ideal membrane material, there have been two fundamental approaches. One was to start off with natural cellulose as the basic polymeric structure. Originally hemodialysis was performed through membranes derived from regenerated, unmodified cellulose. These extremely hydrophilic structures form hydrogels which have a high porosity and small apparent pore sizes. The next membrane generation was prepared by esterification of some of the hydroxyl groups on the cellulose monomer, which trigger complement activation during blood treatment, with acetyl groups

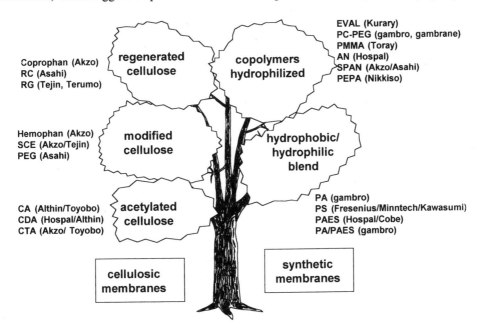

Figure 4. The family tree of dialysis membranes: classes, materials (manufacturers).

or by modification with diethylaminoethyl groups (Hemophan), benzyl groups (SMC), polyethylene glycol or others. The second approach was to use synthetic engineered thermoplastics to create membranes with an improved transport capacity for middle-molecular weight substances and with better biocompatibility properties. The two branches of synthetic membranes represent the two basic techniques for hydrophilization of the materials. One is to co-polymerize polymers during the production process with hydrophilic polymers. This concept is realized for example in the polycarbonate-polyether membranes (Gambrane, Gambro) and the polyacrylonitrile-methallylsulfonate membranes (AN69, Hospal). The second technique involves blending of hydrophobic polymers, e.g. polysulfone (PS), polyarylethersulfone (PAES), and polyamide (PA), with hydrophilic components such as polyethylene glycol (PEG) or polyvinylpyrrolidone (PVP).

The following membrane types, which were developed by the authors, are prepared according to this principle by blending hydrophobic engineering materials with PVP:
1. PA membranes, sterilized by ethylene oxide (Polyflux, Gambro)
2. PAES membranes, sterilized by radiation (Arylane, Hospal/Gambro Group)
3. PA/PAES membranes, sterilized by steam (Polyflux S, Gambro).

The polymer components used have the following repeating chemical structures:

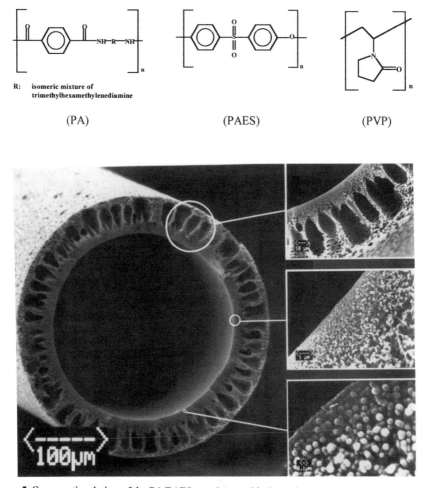

R: isomeric mixture of
     trimethylhexamethylenediamine

(PA)   (PAES)   (PVP)

Figure 5. Cross sectional view of the PA/PAES membrane with the typical asymmetric 3-layer structure.

These membranes are formed by employing a precipitation process, generally referred to as phase inversion. Hereby membranes with highly anisotropic three-layer structures with an extremely thin skin and hydrophilic/hydrophobic microdomains are yielded. In the case of the PA/PAES membrane a three-component polymer blend is used as dope for membrane preparation in order to combine the desirable processing characteristics and bulk properties of PA, PAES and PVP. Partially aromatic amorphous polyamides are known to have a particularly high retention capability for pyrogenic substances.[1,2] With a glass-transition temperature (Tg) of 230 °C PAES has a very high thermal stability compared to other widely used membrane materials such as polysulfone (Tg = 190 °C). Due to the superior mechanical, chemical and thermal material properties solid membrane structures are obtained which can resist chemical reprocessing treatments as well as sterilization procedures with steam at 121 °C without an effect on membrane characteristics. The hydrophilic character of the membrane arises from the addition of PVP. Based on donor/acceptor interactions PVP shows good miscibility with the hydrophobic blend components. It is entrapped in the PA-PAES network and forms an integral part of the polymeric structure, providing the membrane surface with a mosaic-like structure containing hydrophilic microdomains in a hydrophobic base. The mechanical and physical properties of membranes are mainly determined by the basic polymer material, whereas parameters like permeability and blood interaction are predominantly governed by the membrane microstructure and membrane surface. Due to their analogous manufacturing principle which creates similar microstructures and hydrophilic/hydrophobic surface patterns PA, PA/PAES and PAES membranes show comparable permeability and blood-surface interaction behavior.

MEMBRANE STRUCTURE CHARACTERIZATION

As pointed out above, solute transport rates of membranes are mainly governed by their microstructure. The relation between membrane structural properties, such as pore dimensions, porosity, tortuosity, thickness, and the rates for the convective and diffusive transmembrane mass transport in hemodialysis and hemofiltration can be adequately described by the following equations[3]:

$$J_i = \frac{\varepsilon r^2 S C_i^B}{8 \eta \tau} \frac{\Delta P}{\Delta z} \qquad \text{convective transport} \qquad (1)$$

$$J_i = \frac{\varepsilon D_i^M S}{\tau \Delta z} (C_i^B - C_i^B) \qquad \text{diffusive transport} \qquad (2)$$

Here, J_i defines the flow rate of species i through a membrane. The membrane related parameters include the volume fraction ε of the pores, the pore radius r, D_i^M the diffusion coefficient of species i in the membrane, the thickness of the membrane Δz and the membrane tortuosity τ. η represents the dynamic viscosity of the fluid in the membrane and ΔP the hydrostatic pressure difference between the two phases separated by the membrane. C_i^B and C_i^D are the concentrations of the component i on the side facing the bloodstream and the dialysate, respectively.

Based on the equations (1) and (2), one can rewrite the demands on an ideal membrane structure for extracorporeal blood purification as follows:
(1) The pore radius should be within a certain range, (2) the porosity (fraction of the membrane volume which is open to the flow of solvent) of the membrane at the surface as well as in the matrix should be as high as possible to enable high transmembrane fluxes, (3)

the tortuosity (i.e. the measure of the deviation of the structure from cylindrical pores normal to the surface) should be small, (4) the pore size distribution narrow to obtain a sharp molecular weight cutoff curve for the membrane, (5) the diffusion coefficient in the membrane should be high, (6) the smallest diameter of the pores should be on the innermost surface to prevent plugging of the solutes inside the pores, (7) the susceptibility of the membrane to protein adsorption should be limited to prevent narrowing of pores and a declining permeability during blood contact, and finally (8) the active membrane layer (skin) should be as thin as possible, because the permeability is inversely proportional to the thickness of this layer.[4]

Figure 5 shows the scanning electron micrograph of a cross section of the asymmetric PA/PAES membrane. Its structural properties satisfy the considerable demands made on high efficient hemodialysis membranes. The membrane has an integral 3-layer structure in which an ultra-thin skin at the blood contacting surface on the inside of the hollow fiber is supported on large-pore layers. The thickness of the skin layer is about 0.1 μm. It has a high porosity and contains the very fine pores which determine the permeability and solute retention properties of the membrane. The skin is supported by an approximately 10 μm thick sponge-type substructure with high porosity and with increasing pore diameter from the skin towards the bulk structure. A macroporous finger-type structure completes the cross section providing mechanical strength and low mass transport resistance. This strongly asymmetric structure with the pores having the smallest size at the blood contacting surface has lead not only to high solute transport rates and low hydrodynamic flow resistance. It also rejects proteins and therefore prevents the membrane pores from becoming blocked by proteins. As a consequence constant permeation rates can be obtained during application for blood treatment.[5] The microdomain structure showing a distribution of hydrophilic/ hydrophobic moieties of defined dimensions in the skin layer of polymer blend membranes becomes visible by selectively staining hydrophilic PVP with a heavy metal complex and

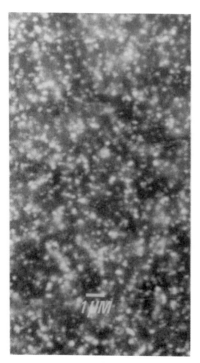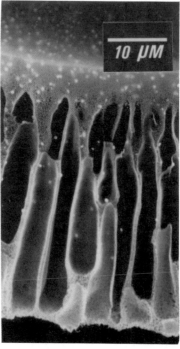

Figure 6. Selectively osmium-stained hydrophilic areas in PA hollow fiber membranes: cross sectional view (right) and top view of the membrane surface (left).

examination by back-scatter electron microscopy. The SEM micrographs (figure 6) reveal the formation of the mosaic-like pattern on the blood-contacting surface of PA membranes. Hydrophilic domains with a size of approximately 50 to 200 nm are homogeneously dispersed in the hydrophobic matrix.[6]

SIEVING PROPERTIES

The degree of openness of a membrane is indicated by the sieving coefficients S for solutes. S is one for solutes that are small in comparison with the pore diameter and becomes zero for solutes that are completely rejected by the membrane. The sieving coefficients can be obtained according to equation (3) by measuring solute concentration in the ultrafiltrate (C^F) and in blood at the inlet and the outlet of the dialyzer (C^{Bi}, C^{Bo}) while operating in hemofiltration mode.

$$S = \frac{2C^F}{(C^{Bi} + C^{Bo})} \tag{3}$$

In the 1980s a possible pathophysiological role of middle-molecular weight substances in the blood of uremic patients was identified. Meanwhile, the role of ß2-microglobulin with a molecular weight of 11,800 D in amyloidosis pathogenesis in chronic uremic patients[7] for example and the effects of systemic cytokines, e.g. TNF-a or IL-6, as well as parts of the complement system, i.e. C3a, in the pathogenesis of septic shock[8] are well established. Therefore, the adequate removal of solutes in this molecular weight spectrum is regarded beneficial in hemodialysis and related treatments. On the other hand, removal of essential functional proteins, e.g. AT III and C1 inhibitor, as well as albumin loss which may contribute to various nutritional and metabolic disturbances in chronic uremic patients should be avoided.

A mean pore radius of approximately six nanometers and a very narrow pore size distribution in the skin layer of the PA/PAES membrane as illustrated in figure 7 result in membrane separation capabilities close to the glomerular barrier in the human kidney. The graph of sieving coefficients for PA/PAES membranes given in figure 8 is characterized by a high permeability for low- and middle-molecular weight substances (sieving coefficient = 1) and a sharp cut-off. With a sieving coefficient for albumin of below 0.001 an effective retention of albumin during hemofiltration procedure is assured.

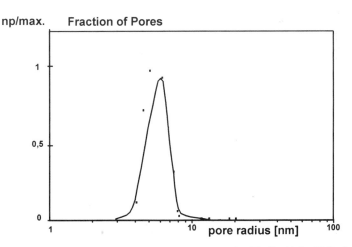

Figure 7. Pore size distribution of PA/PAES membranes determined by liquid-liquid displacement method (curve fit: log. normal distrib.). The highest number of pores can be found at approx. 6 nm.

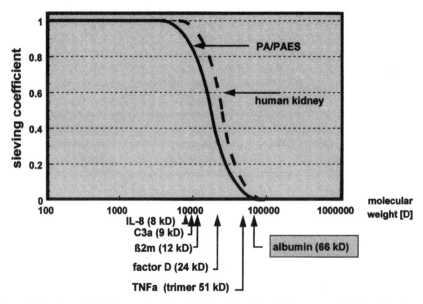

Figure 8. Sieving coefficients as a function of solute molecular weight measured during filtration of polydisperse dextran solutions with PA/PAES membranes.

FILTRATION CHARACTERISTICS

Ultrafiltration is the pressure-driven process by which excess body water is removed from renal failure patients. The linear relation between the ultrafiltration rate (UFR) J_v of a membrane and the driving forces, hydraulic pressure gradient ΔP and osmotic pressure gradient $\Delta \Pi$ across the membrane, is given by equation (4).

$$J_v = Lp \cdot (\Delta P - \Delta \Pi) \tag{4}$$

where Lp is the hydraulic permeability coefficient of the membrane.

The filtrate flux of pure water through a membrane increases linearly with the pressure gradient applied across the dialyzer membrane which is referred to as the transmembrane pressure (TMP). Figure 9 shows typical filtrate flux behavior observed when filtering blood in a PA/PAES high-flux hemodialyzer. The positive intercept on the TMP axis can be explained by the osmotic pressure exerted by plasma proteins that opposes the applied hydrostatic pressure gradient. At low TMP values there is a linear increase of the filtration rate with increasing TMP. At higher TMP values, as a consequence of concentration polarization and the formation of a protein layer at the membrane surface the filtration rate departs from linearity and a pressure-independent plateau is reached. The high surface porosity, the hydrophilicity and the thinness of the skin in the PA/PAES membrane lead to a high ultrafiltration coefficient of 35 mL/h/m²/mmHg which is defined as the slope of the UFR versus TMP in the linear region. The thickness of the polarized layer of protein and blood cells is strongly influenced by the wall shear rate. Therefore, the filtration rate can be increased by increasing blood flow velocities in the hollow fiber membrane as illustrated in figure 9 where the filtration rate is plotted versus the transmembrane pressure at three different blood flow rates. An increase of blood flow rate from 200 ml/min to 400 ml/min leads to a 1.7-fold increase of the plateau value for the flux rate.

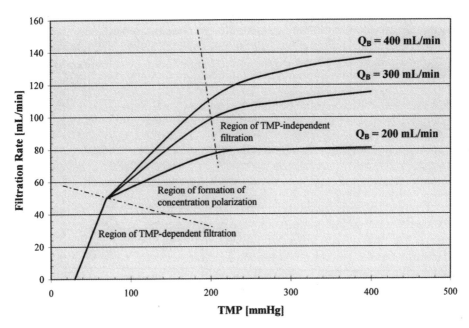

Figure 9. Filtration rate UF of PA/PAES dialyzer as a function of the TMP at different blood flows Q_B and shear rates.

DIFFUSIVE PERMEABILITY

The primary transport mechanism for toxin removal in hemodialysis is solute diffusion across the membrane with a concentration gradient being the driving force for the transport from the blood stream into the dialysate. Factors like high membrane porosity, high membrane mass transfer coefficients, low skin thickness and the hydrophilic character of the membrane yield a high diffusive permeability.

The diffusion coefficient D of molecular components in water decreases roughly in proportion to the cube root of molecular weight. The relation between particle radius r_i and its diffusion coefficient D_i is given by the Stokes-Einstein equation (5):

$$D_i = \frac{RT}{6\pi\eta r_i N_A} \tag{5}$$

Due to the membrane-dependent mechanical resistance diffusion within a membrane is reduced compared to that in free solution. The membrane resistance R_M is the reciprocal of the membrane diffusive permeability P_M, which is related to membrane diffusivity by equation (6):

$$P_M = \frac{D_M}{\Delta z} \tag{6}$$

where Δz is diffusion distance.

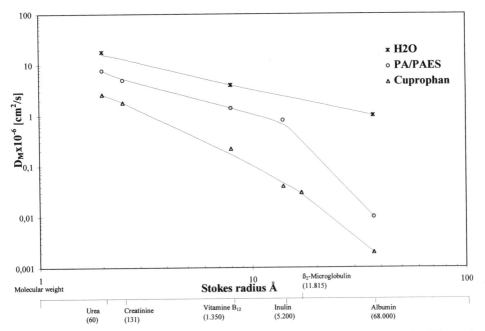

Figure 10. Diffusion coefficients in water and effective membrane diffusion coefficients D_M for different size solutes. D_M calculated with Δz for PA/PAES = 50 μm, for Cuprophan = 16 μm.

Besides membrane resistance also the boundary layers on both the dialysate and blood side contribute to the overall resistance to solute transfer across hemodialysis membranes. The total resistance for the diffusive transport in a hemodialyzer can be expressed as the sum of the resistance of the membrane R_M and the resistances in the laminar boundary layers at the membrane surfaces facing the blood stream R_B and the dialysate R_D:

$$R = R_M + R_B + R_D \tag{7}$$

In hollow fiber dialyzers boundary layer effects on the dialysate side are determined by the dialysate flow distribution, i.e. by the fiber packing and the dialysate flow velocity. Assuming optimal packing the membrane accounts for the principal share of the overall resistance. In high performance dialysis membranes diffusivities for small and middle molecules are close to that in free solution. This is shown in figure 10, where diffusion coefficients of various molecular weight components in water, in regenerated cellulose membranes (Cuprophan) and in PA/PAES membranes are plotted. In order to neglect boundary layer resistances special test conditions were employed for the measurements.[9]

SOLUTE CLEARANCE

The overall mass transfer capacity of a hemodialyzer is expressed in terms of the clearance C_L. It indicates the volume of blood which is completely cleared from a certain substance per unit time (ml/min) and can be calculated by equation (8):

$$C_L = \frac{(C^{Bi} - C^{Bo})Q^{Bi} + Q^F C^{Bo}}{C^{Bi}} \tag{8}$$

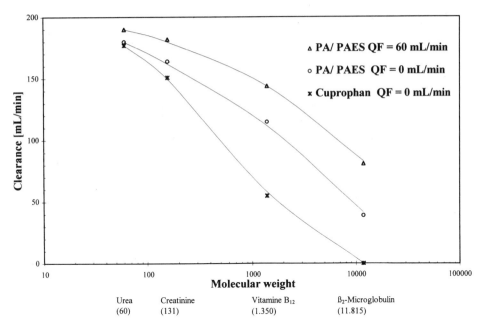

Figure 11. Clearance of PA/PAES dialyzer (Polyflux 14S) compared to a Cuprophan dialyzer of comparable surface. Q_B = 300 mL/min and Q_D = 500 mL/min.

where Q^{Bi} is blood flow rate, Q^F the rate of ultrafiltration, and C^{Bi} and C^{Bo} represent the solution concentrations in blood at the inlet, i, or the outlet, o, of the dialyzer. Clearance increases with flow rates (blood or dialysate), reaching a plateau for high blood flows.

The efficient modern dialyzers produce outstanding clearance performances for low- and middle-molecular weight substances, including phosphate and ß2-microglobulin. This is illustrated in figure 11, which shows clearances measured with a regenerated cellulose (Cuprophan) and a PA/PAES dialyzer (Polyflux S). The clearances were determined for zero net ultrafiltration rate and for an ultrafiltration rate of 60 ml/min. The improved conventional mass transfer caused by the net filtration flux gives a significant additional contribution to the clearance, particularly in the middle-molecular weight spectrum. Efficient and time-constant in vivo elimination of low- and middle-molecular weight substances was reported by Zitta et al.[10] for PA/PAES dialyzers. Solute clearances obtained during clinical hemodialysis can show a significant decrease with treatment time. This behavior is primarily attributable to changes in membrane properties due to protein and/or cell deposition. The magnitude of clearance reduction is a function of several factors, including blood flow rate and membrane type. The hydrophilic asymmetric PA/PAES membranes show no alterations in clearance associated with treatment time.

MEMBRANE BIOCOMPATIBILITY

Ideally, membranes in medical devices should have a very low effect on blood components exposed to artificial surfaces. A number of biological indices could be identified to assess the activation of important reaction pathways in blood. The following mechanisms should show only minimal activation signals: the complement system, the intrinsic pathway of the coagulation system/contact phase, the platelet dependent coagulation system and the stimulation of mononuclear cells. In previous analyses, it could be shown by others and in our laboratory that procoagulatory and immunomodulatory responses as well as reduced activation of mononuclear cells could be best controlled by the design of hydrophilic/hydrophobic microdomain structures.[11, 12, 13] If the domain-size falls below a critical threshold of approximately 100 nm, blood-membrane interaction can be generally minimized and specific activation signals can be efficiently omitted.

Complement and Leucocyte Activation

The complement system is the phylogenetically oldest part of our immune system and is able to differentiate between 'self' and 'non-self' structures on biological and synthetic surfaces. The extent of complement activation is closely related to the expression of activation signals in leucocytes,[14, 15] e.g. activation of neutrophils and increased oxidative stress, upregulation of adhesion receptors, induction of cytokine production (TNF, IL-1). By these activation signals, complement may be involved in endothelial cell dysfunction and tissue reorganization.

The first prerequisite for a non-complement activating membrane is the absence of reactive nucleophilic groups, especially OH-groups, which are able to covalently bind C3-molecules. If there is no stable C3-binding to the surface, complement turnover is controlled by fluid phase inhibitors and no increase of complement specific activation parameters can be measured. Figure 12 shows the relation between white blood cell drop as measured in-vivo[16,17] and the generation of terminal complement complexes by different membrane materials in contact with human plasma. To establish the correlation, we are using a phenomenological correlation between the white blood cell drop of known membranes in clinical use and respective reference membranes in the in-vitro complement assay (e.g. Cuprophan, Hemophan, Polysulfon, AN69). The newly developed PA/PAES membranes showed lowest complement activation index confirming low specific interaction with complement components (e.g. C3).

Low complement activation was also confirmed in clinical studies by Zitta et al.[10] showing a white blood cell drop at 15 min after start of treatment of below 10 % and no significant generation of TCC during the course of dialysis treatment. With this aspect, the microdomain-structured PA/PAES membranes reach highest level of biocompatibility.

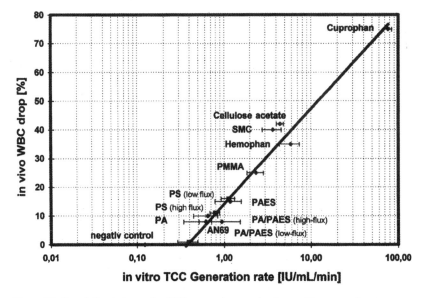

Figure 12. Correlation between the TCC generation rate in vitro and leucocyte drop in vivo.

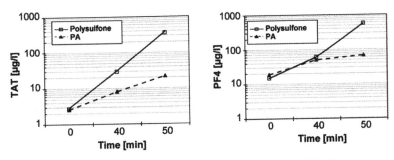

Figure 13. In vitro thrombogenicity evaluation by measurement of TAT and PF4 after exposure to platelet rich plasma: comparison of hydrophilic 3-layer microdomain PA and sponge-type polysulfone membranes.

Reduced Thrombogenicity

There are different elements contributing to reduced thrombogenicity of membranes in extracorporeal circuits. Primarily the activation of the contact phase consisting of factor XIIa, kallikrein and high molecular-weight kininogen has to be strictly avoided for the following reasons: activation of the contact phase system via factor XIIa leads (1) to the release of kallikrein and the induction of bradykinin which is a vasoactive peptide involved in severe dialysis-associated hypotension or hypersensitivity reactions and (2) the initiation of the coagulation cascade, which finally leads to the generation of thrombin and cleavage of fibrinogen and formation of fibrin network.[13]

Contact phase activation is best assessed by in-vitro measurement of kallikrein using specific chromogenic substrates. PA/PAES membranes show no contact phase/kallikrein generation at all, which documents low or even no charged groups at the surface. Basically, these results are not surprising since all polymer components used for membrane formation possess no electrically charged groups.

The formation of fibrin or platelet aggregates can be assessed by in-vitro experiments with platelet-rich plasma. For in-vitro evaluation of different membrane types we are using the formation of thrombin/antithrombin-III (TAT) as indicator of thrombin-activity and the release of platelet factor IV (PF4) as indicator of platelet activation/aggregation. Figure 13 shows significantly lower generation of TAT and PF4 by PA membranes as compared to other synthetic membranes.

In different clinical evaluations[10, 18] it could be shown that platelet counts remain stable during treatment, i.e. changes during treatment remain below 2 %. The low thrombogenic potential of PA/PAES membranes could further be confirmed by Sperschneider et al.[17], showing low thrombin-antithrombin III formation with heparin doses of approximately 5,000 units per treatment (4-5 hrs). As a result of low heparin consumption and limited activation of the coagulation systems, these authors also found a reduced release of van Willebrand-factor as a sign of reduced disturbance of endothelial cell function.

Activation of Mononuclear Cells

Activation of blood components, especially mononuclear cells can be initiated either via the complement system or via direct cell membrane interaction. To assess the compatibility of microdomain structured membranes, we analyzed the induction of 'immediate early genes' by measuring specific C-jun m-RNA for isolated mononuclear cells from human blood or by using the promyelocyte line U937. In figure 14, the induction of 'immediate early genes' as x-fold increase versus non-stimulated control is shown for regenerated cellulose and

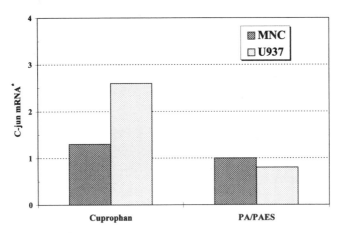

Figure 14. Increase of 'immediate early gene' m-RNA in human mononuclear cells (MNC) and cell-line U937 after incubation with cuprophan and PA/PAES membranes (*x-fold increase compared to non-stimulated control; data are corrected for GAPTH-gene).

microdomain-structured PA/PAES membranes. These data clearly indicate that mononuclear cells become not activated at these surfaces. Further analyses of cell culture supernatant exhibited no increase in cytokine release over negative control.

CONCLUSION

Today synthetic membranes can be adapted to the specific needs required by their clinical application. They offer the possibility to adjust important properties, such as pore size, charge and the balance between hydrophilic and hydrophobic microdomains. The new generation of steam-sterilizable hydrophilic-hydrophobic blend membranes show so far not equaled permeability and biocompatibility. The performance criteria were met by using a highly asymmetric and hydrophilic 3-layer structure and by optimized size and distribution of hydrophilic and hydrophobic domains. The selection of suitable polymer components and process technology steps including steam-sterilization enables an economic large-scale production. In the long term the described profile of polymer blend membranes gives us good reasons to expect an increasing benefit for the patient. A reduction of the occurrence of infections, of the need for hospitalization and medication as well as an improvement of the patient's quality of life will be possible.

References

1. G. Lonnemann, J. Flöge, R. Schindler, T. Behme, B. Lenzner, S. Shaldon, and K.M. Koch, 1989, Passage of cytokine E-coli fragments (CEF) through various hemodialysis (HD) membranes, *Kidney Int.* **35**:354.
2. C.A. Dinarello, G. Lonnemann, R. Maxwell, and S. Shaldon, 1987, Ultrafiltration to reject human interleucin-1-inducing substances derived from bacterial cultures, *J. Clin. Microbiol.* **25**:1233-1238.
3. H. Strathmann, and H. Göhl, 1990, Membranes for blood purification: state of the art and new developments, *Contrib. Nephrol.* **78**:119-141.
4. H. Göhl, R. Buck, and H. Strathmann, 1992, Basic features of the polyamide membranes, i n: The evolution of a synthetic membrane for renal therapy, *Contrib. Nephrol.*, Volume 96 (S. Shaldon, and K.M. Koch, eds.), pp. 1-25, Karger, Basel.

5. J. Flöge, C. Granolleras, G. Deschodt, M. Heck, G. Baudin, B. Branger, O. Tournier, B. Reinhard, G.M. Eisenbach, L.C. Smeby, K.M. Koch, and S. Shaldon, 1989, High-flux synthetic versus cellulosic membranes for ß2-microglobulin removal during hemodialysis, hemodiafiltration and hemofiltration, *Nephrol. Dial. Transplant.* **4**:653-657

6. H. Göhl, C.M. Bell, R. Buck, R. Deppisch, H. Straatman, and M. Pirner, 1992, Visualization and measurement of microdomains in polyamide dialysis membranes: Optimized size facilitates performance and biocompatibility, *Blood Purif.* **10**(2):86.

7. F. Gejyo, S. Odani, T. Yamada, N. Honma, H. Saito, Y. Suzuki, Y. Nakagawa, H. Kobayashi, Y. Maruyama, Y. Hirasawa, M. Suzuki, and M. Arakawa, 1986, ß2-microglobulin: A new form of amyloid protein associated with chronic hemodialysis, *Kidney Int.* **30**:385-390.

8. N. Braun, S. Rosenfeld, M. Giolai, W. Banzhaf, R. Fretschner, H. Warth, C. Weinstock, R. Deppisch, C.M. Erley, G.A. Müller, and T. Risler, 1995, Effect of continuous hemodiafiltration on IL-6, TNF-α, C3a, and patients with SIRS/septic shock using two different membranes, in:. Continuous extracorporeal treatment in multiple organ dysfunction syndrome, *Contrib. Nephrol.*, Volume 116 (H.G. Sieberth, H.K. Stummvoll, and H. Kierdorf, eds.), pp. 89-98, Karger, Basel.

9. E. Klein, F.F. Holland, A. Donnaud, A. Lebeouf, and K. Eberle, 1977, Diffusive and hydraulic permeabilities of commercially available cellulosic hemodialysis films and hollow fibers, *J. Membr. Sci.* **2**:349-364.

10. S. Zitta, A. Mauric, J. Roob, and H. Holzer, 1995, Clinical evaluation of polyflux dialyzer membranes, *Blood Purif.* **13**:28-29.

11. R. Deppisch, M. Betz, G.M. Hänsch, E.W. Rauterberg, and E. Ritz, 1992, Biocompatibility of the polyamide membranes, in: The evolution of a synthetic membrane for renal therapy, *Contrib. Nephrol.*, Volume 96 (S. Shaldon, and K.M. Koch, eds.), pp. 26-46, Karger, Basel.

12. R. Deppisch, E. Ritz, G.M. Hänsch, M. Schöls, and E.W. Rauterberg, 1994, Biocompatibility - perspectives in 1993, *Kidney Int.* **45**(S44):77-84.

13. R. Deppisch, U. Haug, H. Göhl, and E. Ritz, 1994, Role of proteinase/antiproteinase inhibitor disequilibrium in the bioincompatibility induced by artificial surfaces, *Nephrol. Dial. Transplant.* **3**(suppl):17-23.

14. C. Combe, M. Pourtein, V. de Précigout, A. Baquey, D. Morel, L. Potaux, P. Vincendeau, J.H. Bézian, and M. Aparicio, 1994, Granulocyte activation and adhesion molecules during hemodialysis with cuprophane and a high flux biocompatible membrane, *Am. J. Kid. Dis.* **24**(3):437-42.

15. M. Haag-Weber, B. Mai, R. Deppisch, H. Goehl, and W.H. Hoerl, 1994, Studies on biocompatibility of different dialyzer membranes: role of complement system, intracellular calcium and inositol-triphosphate, *Clin. Nephrol.* **41**:245-251.

16. P.R. Craddock, D. Hammerschmidt, J. White, A.P. Dalmasso, and H.S. Jacob, 1977, Complement (C5a)-induced granylocyte aggregation in vitro: a possible mechanism of complement mediated leucostasis and leucopenia, *J. Clin. Invest.* **60**:260-263.

17. L.S. Kaplow, and J.A. Goffinet, 1968, Profound neutropenia during early phase of hemodialysis, *J. Am. Med. Ass.* **203**:1135-1137.

18. H. Sperschneider, G. Steiner, R. Dietrich, R. Deppisch, and G. Stein, 1997, Reduced thrombogenic activity and heparin consumption in extracorporeal blood purification circuits: in-vivo study with microdomain-modified surfaces, *Blood Purif.* **15**(S2):47.

CONTACT AUTHORS

Dr. N. Gürdal ALAEDDİNOĞLU
Middle East Technical University,
Department of Biological Sciences
Biotechnology Research Unit
06531 Ankara / Turkey
Tel: 90 312 210 51 93
Fax: 90 312 210 13 26

Pharm. Betül ARICA
Hacettepe University, Faculty of Pharmacy,
Department of Pharmaceutical Technology,
06100 Sıhhiye, Ankara / Turkey
Tel: 90 312 310 15 24
Fax: 90 312 311 09 06

Dr. Simon BENITA
The Hebrew University of Jerusalem,
School of Pharmacy,
POB 12065
91120 Jerusalem / Israel
Tel: 972 2 675 86 68
Fax: 972 2 534 31 67

Dr. Marie-Ange BENOIT
Universite Catholique de Louvain
Ecole de Pharmacie, Laboratoire de
  Pharmacie Galenique,
Avenue E. Mounier, 73.20-1200
Bruxelles / Belgium
Tel: 32 2 764 73 57
Fax: 32 2 764 73 98

Dr. Nevin ÇELEBİ
Gazi University,
Faculty of Pharmacy,
Department of Pharmaceutical Technology,
  06330 Etiler-Ankara / Turkey
Tel: 90 312 212 79 58
Fax: 90 312 223 50 18

Dr. Y. Murat ELÇİN
Ankara University, Science Faculty,
Chemistry Department,
06100 Tandoğan
Ankara / Turkey
Tel: 90 312 212 67 20
Fax: 90 312 223 23 95

Dr. Ali ERGEN
Hacettepe University,
School of Medicine,
Department of Urology,
06100 Sıhhiye Ankara / Turkey
Tel: 90 312 426 79 59
Fax: 90 312 311 22 62

Dr. Michael GROVES
University of Illinois at Chicago,
Institute for Tuberculosis Research
M/C 964 2014 SEL 950
South Halsted,
Chicago IL 60607
Tel: 1 312 996 39 06
Fax: 1 312 996 46 89

Dr. S. İsmet GÜRHAN
Animal Cell Culture Collection (HÜKÜK),
Foot and Mouth Diseases Institute,
06044 Ankara / Turkey
Tel: 90 312 287 36 00 / 194
Fax: 90 312 287 36 06

Dr. İhsan GÜRSEL
Middle East Technical University,
Department of Biological Sciences
Biotechnology Research Unit
06531 Ankara / Turkey
Tel: 90 312 210 51 93
Fax: 90 312 210 13 26

Dr. Nesrin HASIRCI
Middle East Technical University,
Chemistry Department,
06531 Ankara / Turkey
Tel: 90 312 210 31 93
Fax: 90 312 210 12 80

Dr. Vasıf HASIRCI
Middle East Technical University,
Department of Biological Sciences
Biotechnology Research Unit
06531 Ankara / Turkey
Tel: 90 312 210 51 80
Fax: 90 312 210 13 26

Dr. Feza KORKUSUZ
Middle East Technical University,
Medical Center,
06531 Ankara / Turkey
Tel: 90 312 210 51 80
Fax: 90 312 210 13 26

Dr. Jörg KREUTER
Institut für Pharmazeutische Technologie,
Biozentrum, Marie-Curie-Straße 9,
D-60439 Frankfurt / Germany
Tel: 49 69 798 29 682
Fax: 49 69 798 29 694

Pharm. Erem MEMİŞOĞLU
Hacettepe University, Faculty of Pharmacy,
Department of Pharmaceutical Technology,
  06100 Sıhhiye
Ankara / Turkey
Tel: 90 312 310 15 24
Fax: 90 312 311 09 06

Dr. Filiz ÖNER
Hacettepe University,
Faculty of Pharmacy,
Department of Pharmaceutical
  Biotechnology, 06100 Sıhhiye
Ankara / Turkey
Tel: 90 312 310 15 24
Fax: 90 312 311 09 06

Dr. Sümer PEKER
Ege University,
Engineering Faculty,
Chemical Engineering Department
35100 Bornova-İzmir / Turkey
Tel: 90 232 381 71 43
Fax: 90 232 374 14 01

Dr. Erhan PİŞKİN
Chemical Engineering Department,
Bioengineering Division,
Hacettepe University,
Beytepe, 06530 Ankara / Turkey
Tel: 90 312 299 21 24
Fax: 90 312 440 62 14

Dr. Saime ŞAHİN
Hacettepe University,
Faculty of Dentistry,
Department of Prosthodontics,
06100 Ankara / Turkey
Tel: 90 312 235 18 98
Fax: 90 312 311 37 41

Dr. Duncan STEWART-TULL
Division of Infection &Immunity, IBLS,
Joseph Black Building,
University of Glasgow, Glasgow, G12 8QQ,
Scotland UK
Tel: 141 330 58 38
Fax: 141 330 46 00

Dr. Markus STORR
Gambro Dialysatoren GmBH & Co.KG
Posttach 1323
D-72379 Hechingen / Germany
Tel: 49 7471 17 233
Fax: 49 7471 17 152

Dr. A. Mazhar TOKGÖZOĞLU
Department of Orthopaedics and
  Traumatology, Hacettepe University,
Faculty of Medicine, 06100 Sıhhiye
Ankara / Turkey
Tel: 90 312 468 30 13
Fax: 90 312 466 19 06

Dr. Donald L. WISE
Department of Chemical Engineering,
Center for Biotechnology Engineering,
Northeastern University, Boston, MA 02115
Tel: 617 373 29 92
Fax: 617 373 27 84

M. Hadi ZAREIE
Chemical Engineering Department,
Bioengineering Division,
Hacettepe University,
Beytepe, 06530 Ankara / Turkey
Tel: 90 312 299 21 24
Fax: 90 312 440 62 14

INDEX